Enzyme Active Sites and Their Reaction Mechanisms

Harry Morrison
Department of Chemistry, Purdue University,
West Lafayette, IN, United States

ELSEVIER

ACADEMIC PRESS
An imprint of Elsevier

Academic Press is an imprint of Elsevier
125 London Wall, London EC2Y 5AS, United Kingdom
525 B Street, Suite 1650, San Diego, CA 92101, United States
50 Hampshire Street, 5th Floor, Cambridge, MA 02139, United States
The Boulevard, Langford Lane, Kidlington, Oxford OX5 1GB, United Kingdom

British Library Cataloguing-in-Publication Data
A catalogue record for this book is available from the British Library

Library of Congress Cataloging-in-Publication Data
A catalog record for this book is available from the Library of Congress

ISBN: 978-0-12-821067-3

For Information on all Academic Press publications
visit our website at https://www.elsevier.com/books-and-journals

Publisher: Andre Gerhard Wolff
Acquisitions Editor: Peter B. Linsley
Editorial Project Manager: Samantha Allard
Production Project Manager: Swapna Srinivasan
Cover Designer: Mark Rogers

Typeset by MPS Limited, Chennai, India

Enzyme Active Sites and Their Reaction Mechanisms

Dedication

This book is dedicated to my wife, to my sons and their wives, and to my grandchildren.

Contents

Preface

In a way, this book is the culmination of a journey that began in 1961 in the laboratories of Professor Vlado Prelog, at the Eidgenossiche Technische Hochschule in Zurich. Though Prof. Prelog was the recipient of the Nobel Prize in Chemistry in 1975 for his accomplishments as an organic chemist, it was his lecture at Harvard University in 1960, in which he described his nascent studies of the application of stereochemical principles to the mechanisms of enzymes, that captivated my attention as a graduate student. I never lost my interest in that interface, which eventually became embodied within the subfield known as "bioorganic chemistry." Although my research career at Purdue took a different turn, (the underlying theme of that research has been photochemistry), my studies in photobiology, and the two books that I edited on "Bioorganic Photochemistry," give evidence that the organic chemistry/ biochemistry interface has never been far from my mind.

I have always been fascinated by the capability of an enzyme active site to carry out complex synthetic organic chemistry. One cannot help but be amazed that a small group of amino acids, often with the help of a metal atom or a "cofactor" helper molecule, can do at room temperature and ca. neutral pH what a laboratory chemist might achieve with multiple steps, sophisticated catalysts and much more extreme reaction conditions. Which leads to the obvious question: "why this particular set of 40 enzymes?" The choice of enzymes to include in this volume was made based on a number of factors: (1) the historic role of an enzyme in the evolution of an understanding of enzyme mechanisms, (2) the importance of the enzyme in the field of biochemistry, (3) the novelty of the mechanism, (4) the involvement, and role, of a particular metal or cofactor in the mechanism.

This book is not meant to be a textbook, but rather to serve as a "source" book for scientists at all levels. Typically, each chapter is the distillation of 20 or more papers and book chapters. My goal is that this compilation will well serve the reader's needs, whether it is perused, or is used as an introduction to a particular enzyme or process.

Harry Morrison

Purdue University, West Lafayette, IN, United States

Acknowledgments

First and foremost, I thank my wife, Harriet, for her patience and support throughout this project. This book could never have been written without that support. I also want to thank my secretary, Ann Cripe, who painstakingly created all of the original images in this book. Finally, my thanks to Dr. Minou Bina, with whom I discussed the book concept as it germinated, and who encouraged me to bring it to fruition.

I also want to express my gratitude to the Purdue University Chemistry Department for providing the resources that allowed me to pursue this project. Likewise, the Florida Atlantic University Chemistry Department has been most generous in periodically hosting me during the writing of this book.

Chapter 1

Acetylcholinesterase

1.1 Acetylcholinesterase

Acetylcholinesterase (EC 3.1.1.7; AChE; acetylcholine hydrolase) is a serine hydrolase that catalyzes the hydrolysis of acetylcholine (ACh)—a neurotransmitter (Fig. 1.1). As such it regulates the concentration of ACh at the

$$\underset{\text{Acetylcholine}}{CH_3\overset{\overset{\displaystyle O}{\|}}{C}OCH_2CH_2\overset{+}{N}(CH_3)_3} \xrightarrow{H_2O} CH_3CO_2H + \underset{\text{Choline}}{HOCH_2CH_2\overset{+}{N}(CH_3)_3}$$

FIGURE 1.1 The overall chemistry catalyzed by acetylcholinesterase.

synapse. The complete blockage of this enzyme (e.g., by the nerve gas sarin) is lethal. It is an exceedingly rapid enzyme, the rate of which approaches diffusion control.

1.2 Physiological function

AChE is found in all types of conducting tissue in animals. It terminates impulse transmission by the hydrolysis of ACh.

1.3 Key structural features

The enzyme's active site has two primary binding subsites—one (esteratic site) in which hydrolysis occurs and a second (anionic site) in which the quaternary ammonium group of ACh is bound. There is also a peripheral binding site at the entrance to the gorge, which provides the first point of contact for substrates and is a binding site for some inhibitors. The structural details derive heavily from the X-ray analysis of the enzyme isolated from the "electric ray" *Torpedo californica*. These indicate that AChE is different from other more common serine hydrolases (e.g., α-chymotrypsin) in that the active site contains a catalytic triad, (Ser_{200}—His_{440}—Glu_{327}), that contains glutamate in place of the more common aspartate amino acid (AA). The triad lies 4 Å from the bottom of an "aromatic gorge" containing 14 aromatic AAs. One of these, $Tryp_{84}$, is a key component of the "anionic site";

Enzyme Active Sites and their Reaction Mechanisms. DOI: https://doi.org/10.1016/B978-0-12-821067-3.00001-5

the pi electrons of its indole ring function as a Lewis base and interact strongly with the ammonium group. This interaction with $Tryp_{84}$ is of sufficient import that it directs ACh to adopt an extended conformation within the binding pocket rather than the *gauche* conformation it otherwise adopts in solution. The acyl group sits in a binding pocket consisting of Phe_{288}, Phe_{290}, and His_{440}. The tetrahedral oxyanion intermediate hydrogen bonds to Gly_{118}, Gly_{119}, and Ala_{201}, the constituents of its "oxyanion hole" (Fig. 1.2).

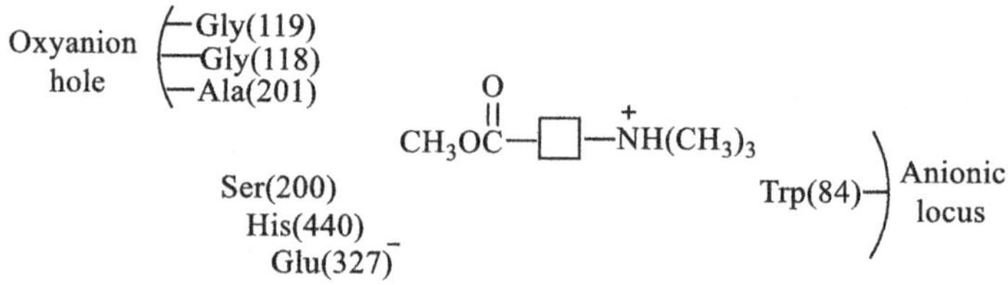

FIGURE 1.2 Schematic rendering of the active site for acetylcholinesterase.

1.4 Reaction sequence

The hydrolytic sequence involves initial nucleophilic attack by the Ser_{200}—OH on the acetyl carbonyl group to form a tetrahedral "oxyanion" intermediate typical of carboxylic ester exchange reactions (alcoholysis). This intermediate fragments to form choline and an acyl-enzyme ester (acetyl-AChE). Hydrolysis of the acyl-enzyme then releases acetate and regenerates the AChE Ser_{200}—OH. The general reaction sequence for alcoholysis of an ester is outlined in Fig. 1.3.

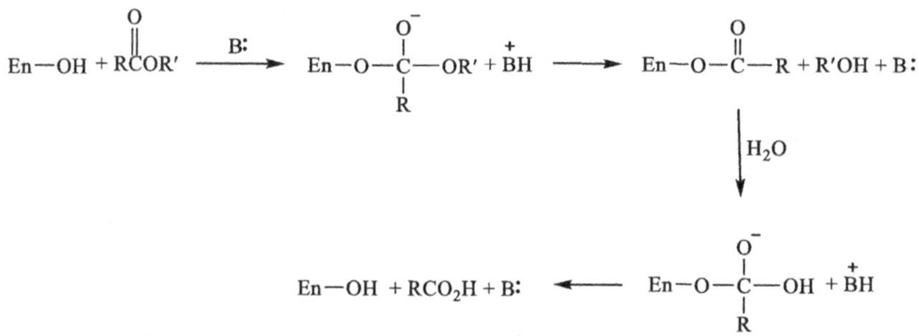

FIGURE 1.3 General reaction sequence for the AChE-catalyzed hydrolysis of acetylcholine via an initial ester exchange reaction. *AChE*, Acetylcholinesterase.

1.5 Mechanism and the role of active site residues

The multistep sequence is outlined in Fig. 1.4. In step (a), ACh is bound into the active site. Though a number of AAs contribute to the binding at this

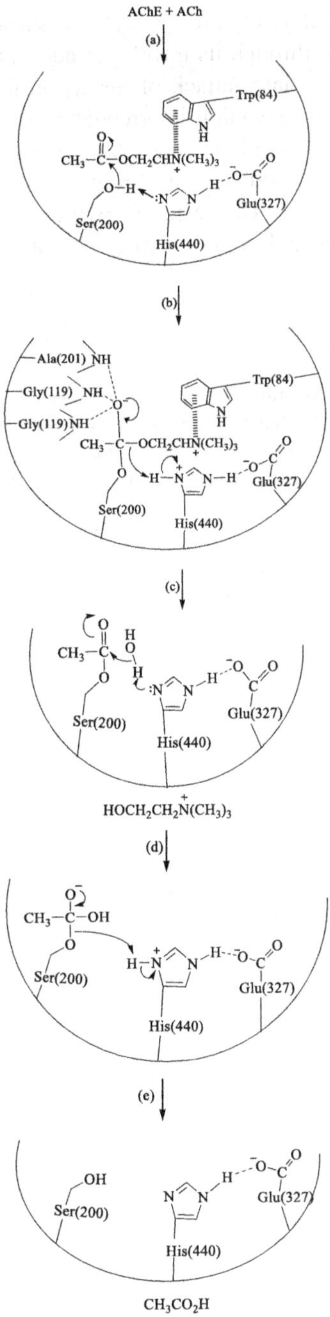

FIGURE 1.4 Detailed mechanism for the conversion of acetylcholine to choline showing the role of key AAs at the active site. *AA*, Amino acid.

point, especially in the acyl binding pocket, we show only Trp_{84} interacting with the choline moiety through its novel interaction as a Lewis base. In step (b), the catalytic triad initiates attack of Ser_{200} on the ester carbonyl group, resulting in a tetrahedral, oxyanion intermediate. The AAs that stabilize this intermediate in the "oxyanion hole" are Ala_{201}, Gly_{118}, and Gly_{119}. The catalytic triad facilitates the expulsion of choline in step (c), with concomitant formation of acetyl-AChE. This ester is hydrolyzed in steps (d) and (e), again by deprotonation and re-protonation chemistry involving His_{440} and Glu_{327}.

Leading references

Dvir, H.; Silman, I.; Harel, M.; Rosenberry, T. L.; Sussman, J. L. *Chem.-Biol. Interact.* **187**, 10–22 (2010); Shafferman, A.; Barak, D.; Stein, D.; Kronman, C.; Velan, B.; Greig, N. H.; Ordentlich, A. *Chem.-Biol. Interact.* **175**, 166–172 (2008); Tõugu, V. *Curr. Med. Chem.-Central Nervous System Agents* **1**, 155–170 (2001); Quinn, D. M. *Chem. Rev.* **87**, 955–979 (1987).

Chapter 2

Aconitase

2.1 Aconitase

Aconitase (EC 4.2.1.3) is the second of a series of enzymes involved in the citric acid cycle. The citric acid cycle is also known as the tricarboxylic acid cycle and the Krebs cycle. Aconitase's role is to isomerize citric acid to isocitric acid (Fig. 2.1).

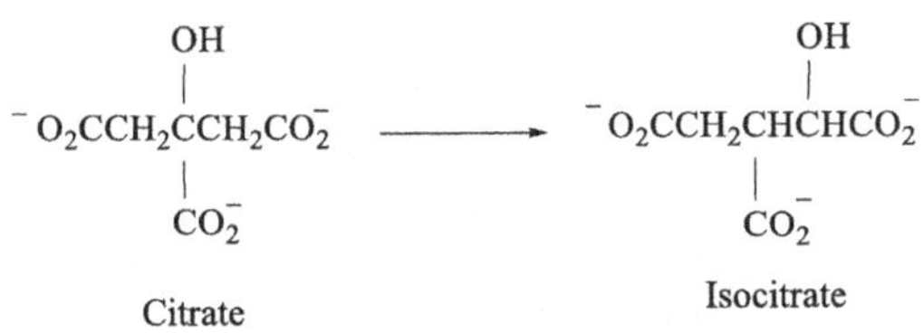

FIGURE 2.1 The overall chemistry catalyzed by aconitase.

2.2 Physiological function

Two aconitase enzymes have been isolated from mammalian tissue, mitochondrial (m-) and cytoplasmic (c-). The c-enzyme is the more stable of the two and is typically isolated from the liver. c-Aconitase has molecular weight of 98,400 Da and 889 AA residues. Crystallography and mutational studies indicate that all the residues that are considered essential to the active site are conserved between the two enzymes. The role of this enzyme is to isomerize citrate to isocitrate. *cis*-Aconitate is formed as an intermediate in this transformation.

2.3 Key structural features

This enzyme incorporates an iron–sulfur, $[4Fe{-}4S]^{2+}$, cluster as a key component of its active site. The cluster is cubic with an Fe^{2+} atom at one position. All of the Fe atoms have a tetrahedral geometry. Three have cysteine-derived ligands, which have been identified as Cys_{358}, Cys_{421}, and Cys_{424}. The fourth Fe atom (Fe_a below) is bound to a hydroxyl group in the absence of substrate. The cluster is depicted in Fig. 2.2. As seen in Fig. 2.4, Fe_a becomes hexa-coordinate and takes on an octahedral geometry when bound to the substrate.

Enzyme Active Sites and their Reaction Mechanisms. DOI: https://doi.org/10.1016/B978-0-12-821067-3.00002-7

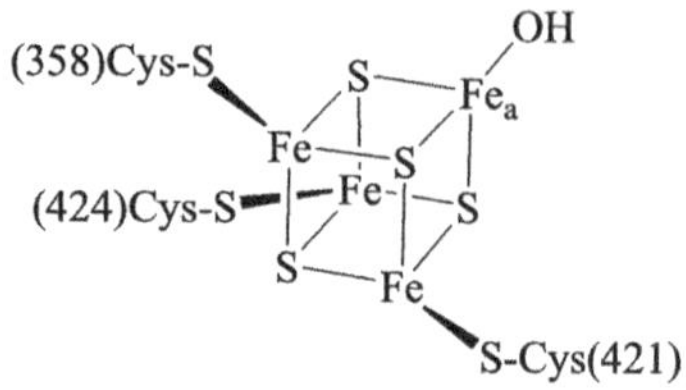

FIGURE 2.2 The $[4Fe{-}4S]^{2+}$ cluster in the aconitase active site.

Twenty-one AA residues have been identified as associated with the active site. Those identified as having particular mechanistic importance are Asp_{100}, His_{101}, His_{167}, Glu_{262}, Arg_{580}, and Ser_{642}.

2.4 Reaction sequence

This enzyme catalyzes a two-step reaction that converts citrate to isocitrate through the intermediacy of *cis*-aconitate. The first step is a dehydration to the alkene involving removal of the tertiary alcohol group from the 2 position of the propane chain. In the second step rehydration of the alkene occurs, with the hydroxyl group added to the (less-substituted) 1 position to form a secondary alcohol. This second step is an anti-Markovnikov addition reaction and thus thermodynamically disfavored. Citrate, at 91%, therefore dominates when the reaction is at equilibrium. There is evidence that the hydrogen atom removed in the first step, and added in the second step, is one and the same. The conversion of a tertiary alcohol to a secondary alcohol is essential to the subsequent oxidative chemistry in the metabolic cycle. The reaction sequence is shown in Fig. 2.3.

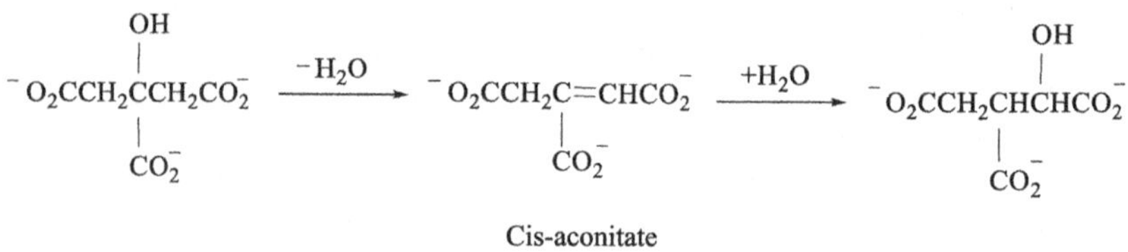

FIGURE 2.3 General reaction mechanism for the conversion of citrate to isocitrate.

2.5 Detailed mechanism and the role of the active site residues

The sequence, based on crystal structures from m-aconitase, is shown in Fig. 2.4. The reaction is initiated by protonation of the hydroxyl group bound to the iron of the Fe–S cluster by His_{167}, facilitated by Glu_{262} (step (a)). Citrate then binds to the modified cluster [step (b); the representation of the cluster has been reduced to the key iron component]. A more complete image shows the

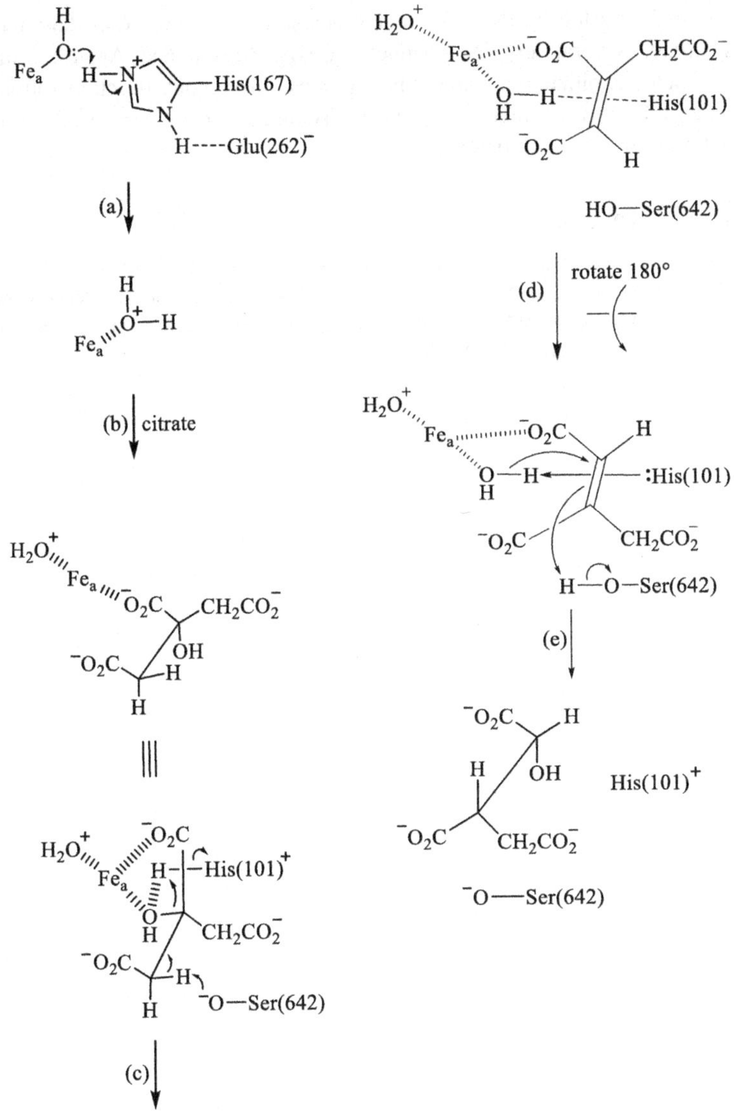

FIGURE 2.4 Detailed mechanism for the m-aconitase-catalyzed conversion of citrate to isocitrate showing the involvement of key amino acids at the active site.

stereochemistry of the citrate additionally complexed with the iron and with key amino acids at the active site. In step (c), the combination of base-catalyzed deprotonation by Ser_{642}, and protonation of the departing hydroxyl group by His(101) (facilitated by Asp_{100}) leads to *cis*-aconitate. A novel reorientation of the *cis*-aconitate, either within the active site or by extrusion and rebinding of the substrate, is shown in step (d). This is followed by stereospecific hydration of the C=C, assisted by the same amino acids, to form the 2*R*,3*S* diastereomer of isocitrate. Several arginine residues (Arg_{447}, Arg_{544}, and Arg_{580}; data not shown) help stabilize the carboxylate groups throughout the transformation. Note that all steps are reversible, that is, isocitrate can be converted to citrate through the same general pathway.

Leading references

Lloyd, S. J.; Lauble, H.; Prasad, G. S.; Stout, C. D. *Protein Sci.* **8**, 2655–2662 (1999); Beinert, H.; Kennedy, C. K.; Stout, C. D. *Chem. Rev.* **96**, 2335–2373 (1996); Lauble, H.; Kennedy, M. C.; Beinert, H. *J. Mol. Biol.* **237**, 437–451 (1994); Lauble, H.; Kennedy, M. C.; Beinert, H.; Stout, C. D. *Biochemistry* **31**, 2735–2748 (1992).

Chapter 3

Adenosine deaminase

3.1 Adenosine deaminase (adenosine aminohydrolase)

Adenosine deaminase (adenosine aminohydrolase) (EC 3.5.4.4; ADA) is one of the enzymes responsible for nucleotide and nucleoside degradation. It hydrolyzes adenosine and 2′-deoxyadenosine to inosine and 2′-deoxyinosine, respectively, with the concomitant release of ammonia. As such, it is part of a family of purine/pyrimidine deaminase enzymes. The overall reaction is shown in Fig. 3.1.

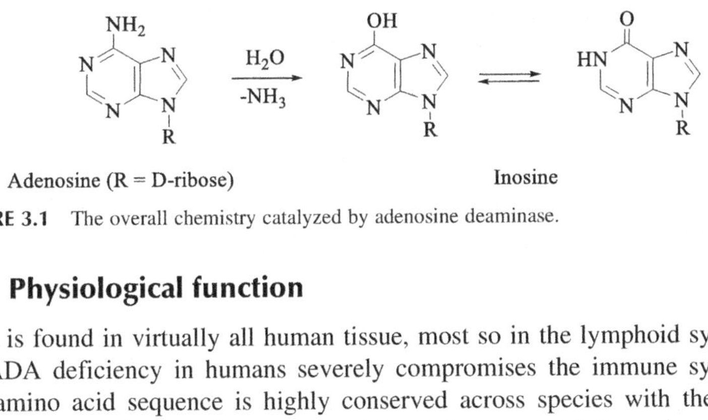

FIGURE 3.1 The overall chemistry catalyzed by adenosine deaminase.

3.2 Physiological function

ADA is found in virtually all human tissue, most so in the lymphoid system. An ADA deficiency in humans severely compromises the immune system. The amino acid sequence is highly conserved across species with the first X-ray structure derived from murine ADA. More recent information about the active site was obtained with the X-ray analysis of the enzyme bound to 6-hydroxyl-1,6-dihydropurine ribonucleotide, a "reaction coordinate" analog that mimics an intermediate hydroxyl addition product.

3.3 Key structural features

ADA is a metalloenzyme with the active site consisting of a zinc ion attached to three His residues (His_{15}, His_{17}, and His_{214}), Asp_{295}, and a water molecule. Three of the Zn-bound His residues are stabilized by hydrogen-bonding to nearby AAs-His_{15} to Glu_{260}, His_{214} to Asp_{181}, and Asp_{295} to Ser_{265}. The fourth residue, His_{17}, interacts with the 5′-OH of the substrate. Fig. 3.2 is a schematic rendering of the active site into which has been bound an analog, 2-deazaadenosine.

Enzyme Active Sites and their Reaction Mechanisms. DOI: https://doi.org/10.1016/B978-0-12-821067-3.00003-9

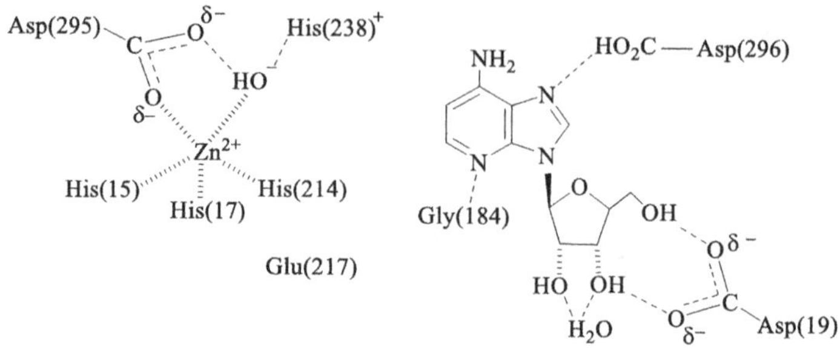

FIGURE 3.2 A schematic rendering of the active site for adenosine deaminase containing 2-deazaadenosine. *Adapted with permission from Wilson, D. K. and Quiocho, F. A.* Biochemistry *32, 1689–1694 (1993). Copyright 1993, American Chemical Society.*

3.4 Reaction sequence

The hydrolysis chemistry is generally accepted to involve the nucleophilic addition of a zinc-activated water molecule to the adenosine C6 atom, followed by the elimination of ammonia. The general reaction resembles the hydrolysis of an amidine, as shown in Fig. 3.3. Note that such reactions typically require an initial protonation by mild acid catalysis to assist the addition reaction, as well as protonation of the amino leaving group prior to its expulsion as ammonia.

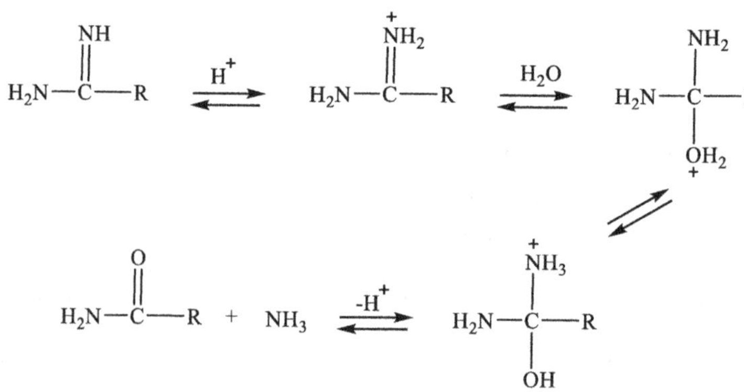

FIGURE 3.3 Acid-catalyzed hydrolysis of an amidine.

3.5 Detailed mechanism and the role of active site residues

The "classic" mechanism, dating back to 1993, envisages His_{238} acting as a base to deprotonate the Zn-bound water with concomitant attack of OH on the purine ring at C6. The sequence is shown in Fig. 3.4. In step (a), the Glu_{217} acts as a proton donor to N1 as addition of OH takes place stereospecifically on the B face of the ring to form the 6(R) intermediate. The location of the OH is guided by the orientation of Asp_{295}. In step (b), Glu_{217} is now the base and His_{238} is the proton donor. Ammonia and inosine (as its enol tautomer) are released.

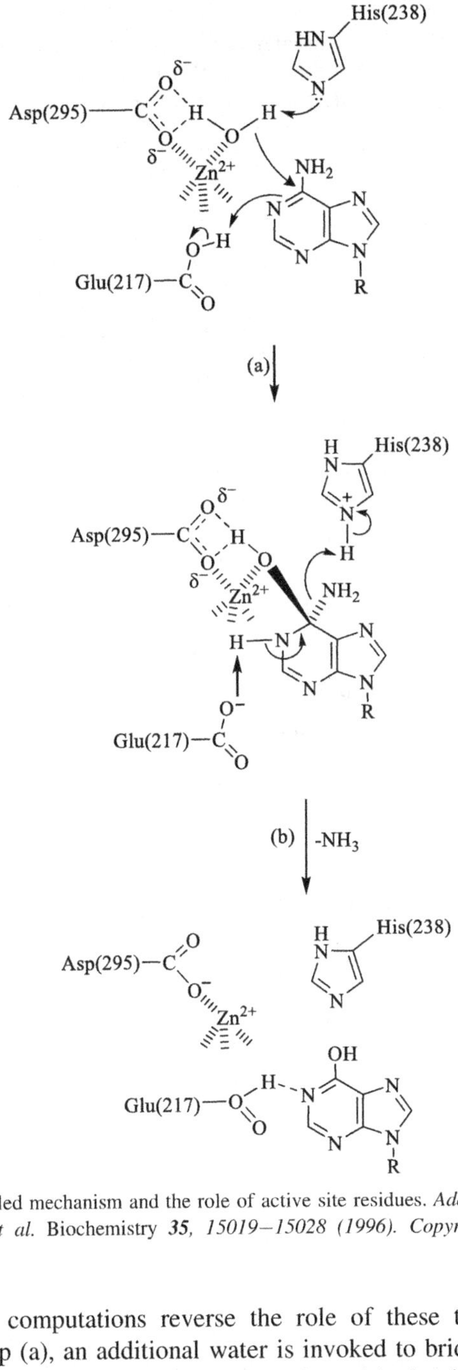

FIGURE 3.4 Detailed mechanism and the role of active site residues. *Adapted with permission from Sideraki, V. et al.* Biochemistry ***35**, 15019–15028 (1996). Copyright 1996, American Chemical Society.*

More recent computations reverse the role of these two key residues (Fig. 3.5). In step (a), an additional water is invoked to bridge the Zn-bound water and glutamate base to generate a Zn-bound hydroxide anion. This is

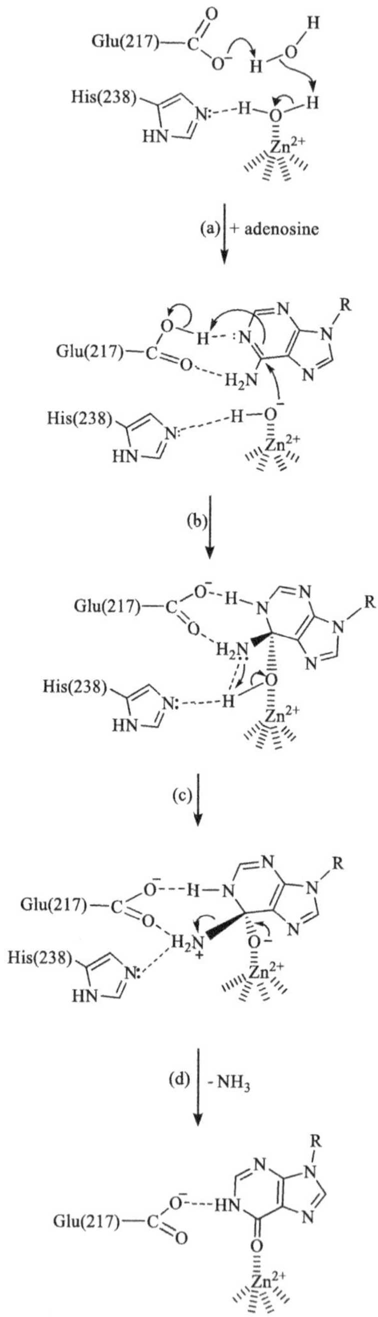

FIGURE 3.5 Alternative mechanism for adenosine deaminase. *Adapted with permission from Wu, J. et al.* Comp. Chem. ***31****, 2238–2247 (2010). Copyright 2010 John Wiley and Sons.*

followed by nucleophilic attack on the adenosine ring [step (b)]. The O—H proton is transferred to the amino group in step (c) and inosine is formed and ammonia expelled in step (d).

Several options for the transition state for step (c) have been compared. The most favorable is shown in Fig. 3.6. The imidazole is said to "mediate" the transfer from the oxygen to the adenosine nitrogen without ever forming a full His(238) N—H bond.

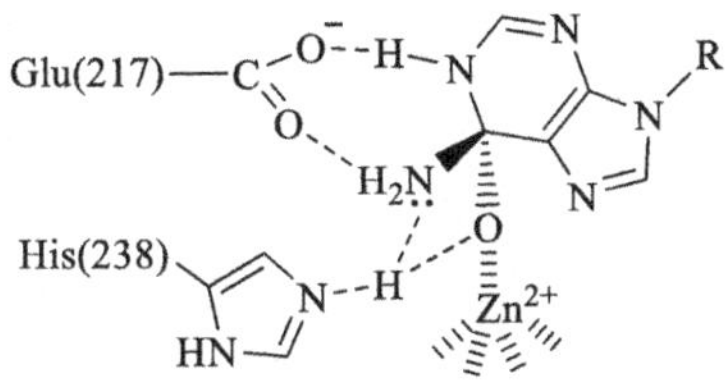

FIGURE 3.6 Calculated transition state for step (c) of Fig. 3.5. *Copyright 2010 John Wiley and Sons. See Fig. 3.5.*

Leading references

Cortés, A.; Gracia, E.; Moreno, E.; Mallol, J.; Lluis, C.; Canela, E. I.; Casadó, V. *Med. Res. Rev.* **35**, 85–125 (2015); Wu, X.-H.; Zou, G.-L.; Quan, J.-M.; Wu, Y.-D. *J. Comp. Chem.* **31**, 2238–2247 (2010); Luo, M.; Singh, V.; Taylor, E. A.; Scramm, V. L. *J. Am. Chem. Soc.* **129**, 8008–8013 (2007); Sideraki, V.; Wilson, D. K.; Kurz, L. C.; Quiocho, F. A.; Rudolph, F. B. *Biochemistry* **35**, 15019–15028 (1996); Wilson, D. K. and Quiocho, F. A. *Biochemistry* **32**, 1689–1694 (1993).

Chapter 4

Alcohol dehydrogenase (horse liver)

4.1 Horse liver alcohol dehydrogenase

Horse liver alcohol dehydrogenase (EC 1.1.1.1; ADH1E; horse LADH; horse ADH) is a member of the "class I" subfamily of a superfamily of zinc-containing, "medium chain", alcohol dehydrogenases (ADH). Humans have seven groups of genes for ADH: ADH1 genes code for "class I" isoenzymes; ADH2 for class II isoenzymes, etc. The class I isoenzymes have been of particular interest because of their role in detoxifying biogenic and dietary alcohols, particularly ethanol. In humans, this function falls to the ADH1B*1 isoenzyme, an excellent and well-characterized model for which is horse ADH. This zinc metalloenzyme (see also Chapter 8: Carboxypeptidase A; zinc is the second most abundant metal in humans) reversibly oxidizes alcohol to aldehydes and ketones. It utilizes NAD^+ as a cofactor (see Chapter 14: Dihydrolipoamide Dehydrogenase for the structures of NAD^+ and NADH) (Fig. 4.1).

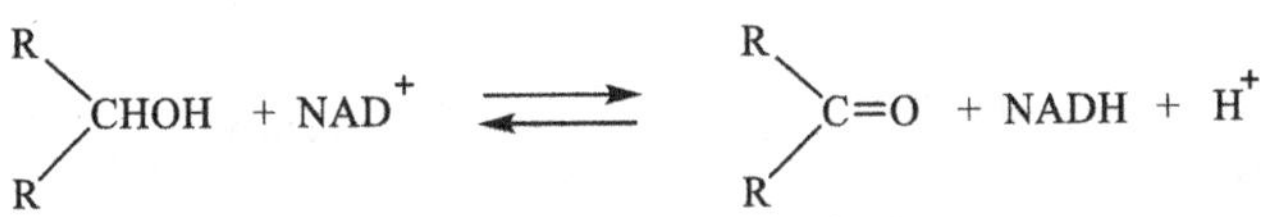

Forms ketones and aldehydes (one R = H)

FIGURE 4.1 The overall chemistry catalyzed by alcohol dehydrogenase.

4.2 Physiological function

As noted earlier, the classic example of ADH activity is the conversion of ethanol to acetaldehyde, but other analogous oxidations are also of import. The oxidation is reversible and, for example, yeast and bacteria extensively utilize this enzyme to carry out reductions. Human ADH1B*1 is primarily found in the liver.

4.3 Key structural features

X-ray structures for the horse liver apoenzyme, the enzyme/NAD^+ holoenzyme, and a ternary complex with the substrate are available. Horse ADH is

Enzyme Active Sites and their Reaction Mechanisms. DOI: https://doi.org/10.1016/B978-0-12-821067-3.00004-0

a dimer of two identical chains. Both chains contain a Zn^{II} atom but only one chain has a binding site for the coenzyme. Therefore only one Zn^{II} (the "catalytic zinc") is involved in the catalytic oxidation reaction. The other zinc atom (structural zinc) is critical for the structural stability of the protein. The coenzyme and catalytic sites exist in separate domains, with access to the active site via a cylinder that extends c.15 Å from the catalytic Zn to the enzyme surface. The NAD^+ binds first, followed by the substrate. The ternary complex has a different (closed) conformation than does the apoenzyme (open), the transition between these occurring upon binding of the coenzyme. The two domains rotate toward one another in the transition, bringing the coenzyme and substrate closer to one another. The active site containing benzyl alcohol as a substrate is shown in Fig. 4.2.

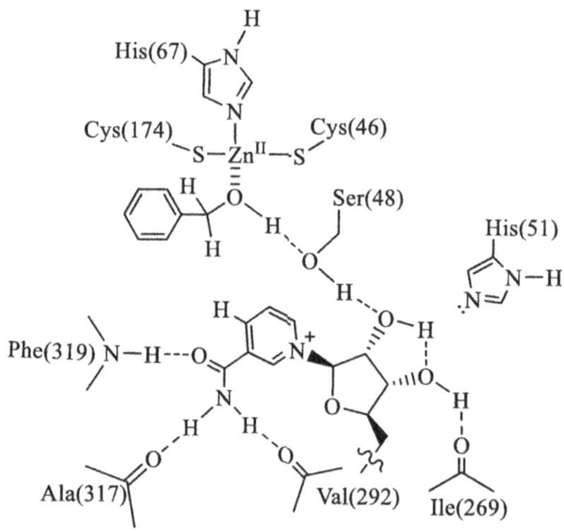

FIGURE 4.2 The active site of LADH-containing benzyl alcohol. *LADH*, Horse liver alcohol dehydrogenase. *Adapted with permission from Agarwal, P.K. et al.* J. Am. Chem. Soc. ***122***, *4803–4812 (2000). Copyright 2000, American Chemical Society.*

4.4 Reaction sequence

The reaction is summarized in Fig. 4.3. As will be seen later, the transfer of hydrogen from the alkoxide anion to the NAD^+ $^+$occurs through the

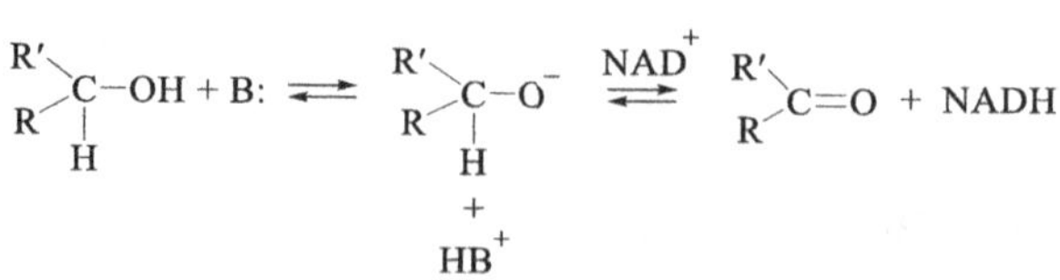

FIGURE 4.3 General mechanism for the base-catalyzed oxidation of a secondary alcohol to a ketone.

transfer of a hydride anion. The acidity of the OH groups in the alcohol substrates is increased by as much as 8–9 pKa units through binding to the zinc ion.

4.5 Detailed mechanism and the role of the active site residues

A detailed description of the mechanism is presented in Fig. 4.4. Step (a) involves the binding of NAD^+ to the apoenzyme. Only the key residues involved with the nicotinamide portion of the coenzyme are shown. As with carboxypeptidase A, the zinc ion is bound to the enzyme through 3 AAs with water as the fourth ligand. In step (b), the water is displaced by the substrate (here, benzyl alcohol). Though not detailed here, it has been suggested that Glu_{68} may initially displace the water/hydroxide from the Zn and, in turn, be displaced by the substrate. Step (c) involves deprotonation of the alcohol (via a chain of proton transfers) to create a zinc-coordinated alkoxide anion. The interaction between the alkoxide anion and the hydrogen on Ser_{48} has been characterized as a "low-barrier" H-bond (see also Chapter 9: Chymotrypsin and Chapter 10: Citrate Synthase). In step (d), the alkoxide anion transfers a hydride ion to the NAD^+. The hydride transfer reaction is stereospecific, with the proR hydrogen atom preferentially transferred to the *re* face of the nicotinamide ring. Val_{203} (not shown) lies nearby the nicotinamide ring and its steric bulk facilitates the hydride transfer by moving the hydride-accepting nicotinamide carbon closer to the substrate C − H bond. It also induces the ring to adopt a more reactive (puckered, quasi-boat) conformation. The hydride donor and acceptor atoms are spatially quite close; isotope effect studies and quantum mechanical calculations indicate that the transfer utilizes a "tunneling" mechanism (see Chapter 24: Methylmalonyl Coenzyme A Mutase). Water displaces the product carbonyl compound from the zinc atom and the NADH is released.

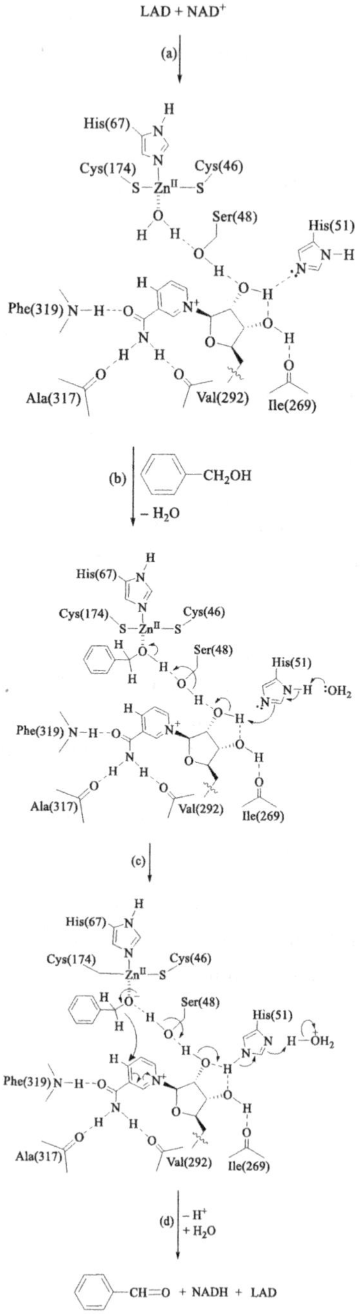

FIGURE 4.4 Detailed mechanism for the alcohol dehydrogenase-catalyzed oxidation of a primary alcohol showing the involvement of key amino acids at the active site.

Leading references

Kim, K. and Plapp, B. V. *Chem-Biol. Inter.* **302**, 172–182 (2019); Shanmuganatham, K. K.; Wallace, R. S.; Lee, A. T.-I.; Plapp, B. V. *Protein Sci.* **27**, 750–768 (2018); Kim, Y. H.; Gogerty, D. S.; Plapp, B. V. *Arch. Biochem. Biophys.* **653**, 95–106 (2018); Plapp, B. V.; Savarimuthu, B. R.; Ferraro, D. J.; Rubach, J. K.; Brown, E. N. *Biochemistry* **56**, 3632–3646 (2017); Jornvall, H.; Landreh, M.; Östberg, L. J. *Chem.-Biol. Interact.* **234**, 75–79 (2015); Plapp, B. V. *Arch. Biochem. Biophys.* **493**, 3–12 (2010); Luo, J. and Bruice, T. C. *Biophys. Chem.* **26**, 80–85 (2007); Hammes-Schiffer, S. and Benkovic, S. F. *Annu. Rev. Biochem.* **75**, 519–541 (2006); Hayward, S. and Kitao, A. *Biophys. J.* **91**, 1823–1831 (2006); Gibbons, B. J. and Hurley, T. D. *Biochemistry* **43**, 12555–12562 (2004); Agarwal, P. K.; Webb, S. P.; Hammes-Schiffer, S. *J. Am. Chem. Soc.* **122**, 4803–4812 (2000); Pettersson, G. and Klinman, J. P. *Crit. Rev. Biochem.* **21**, 349–383 (1987).

Chapter 5

Aldehyde dehydrogenase

5.1 Aldehyde dehydrogenase

Aldehyde dehydrogenase (EC 1.2.1.3; ALDH) is a superfamily of 19 isozymes that oxidize aldehydes to carboxylic acids, most commonly employing NAD^+ (several use $NAD(P)^+$) as a cofactor. Some may also involve a Mg^{2+} ion. Of the ALDH isozymes, several of the ALDH1 family (ALDH1A1, ALDH1A2, and ALDH1A3) and ALDH2 have been the most intensively studied because of their importance in the formation of retinoic acid from retinal, the oxidation product of retinol (vitamin A) (Fig. 5.1), and the conversion of acetaldehyde to acetic acid (the second step in the metabolism of ethanol). The overall reaction is shown in Fig. 5.2.

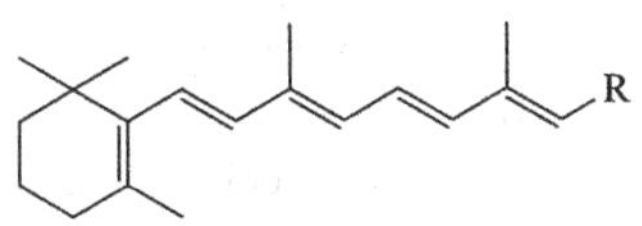

Retinol	$R = CH_2OH$
Retinal	$R = CH{=}O$
Retinoic acid	$R = CO_2H$

FIGURE 5.1 The structures of retinol, retinal, and retinoic acid.

FIGURE 5.2 Overall reaction catalyzed by aldehyde dehydrogenase.

5.2 Physiological function

ALDH1A1, ALDH1A2, and ALDH1A3 are cytosolic enzymes that catalyze the conversion of retinal to retinoic acid. The biosynthesis of retinoic acid is

Enzyme Active Sites and their Reaction Mechanisms. DOI: https://doi.org/10.1016/B978-0-12-821067-3.00005-2

associated with embryogenesis and fetal development. Mutations of ALDH enzymes are known to cause several neurological diseases. ALDH1A1 and (mitochondrial) ALDH2 are also important in the metabolic destruction of ethanol by efficiently converting acetaldehyde to acetic acid. A single base-pair mutation in ALDH2 (Glu_{487} to Lys_{487}), commonly found in East Asians, inactivates this enzyme. The result is an accumulation of acetaldehyde and the so-called "facial flushing" syndrome (as well as an increased risk of esophageal cancer) when alcohol is consumed by these individuals. These enzymes are found in the highest concentration in the liver, but ALDH1A1 also helps to maintain corneal transparency. ALDHs are important in detoxifying exogenous aldehydes which, as a functional group family, are highly reactive with several biochemical entities.

5.3 Key structural features

ALDH1A1, ALDH1A2, and ALDH2 are homotetramers (also described as a dimer of dimers), with each subunit consisting of three domains: catalytic, cofactor binding, and oligomerization (bridging). ALDH1A3 is a homodimer. The active site is heavily conserved among these isozymes. It lies between the cofactor and catalytic domains, with access provided by a funnel which is formed at the interface of the three domains. The nature of the residues lining this funnel determines the enzyme's aldehyde specificity. Divalent metal cations (e.g., Mg^{2+}, Ca^{2+}) enhance enzyme activity, presumably by stabilizing the binding of the NAD^+ cofactor. Binding of NAD^+ precedes, and in fact facilitates, the binding of the aldehyde substrate. The discussion below will emphasize the structural information available for ALDH2.

Two conformations of the NAD^+-binding pocket have been identified and labeled as the "hydride transfer" (aka "closed" and "extended") and "hydrolysis" ("open" and contracted") conformations (see mechanism below). The hydrogen-bonding network that makes up the "hydride transfer" binding pocket is shown in Fig. 5.3. Particularly noteworthy are the placement of the magnesium ion (and the adjacent structured water molecule) and the location of Cys_{302} (see mechanism below). It has been proposed that the tight binding of the magnesium ion with the NADH enhances the rate of conversion of the closed to the open conformation after reduction of the cofactor has occurred. Thr_{244} and Glu_{487} (both not shown) have been identified as conserved residues that are important for enzyme function. Thr_{244} helps to position the nicotinamide ring for hydride transfer (see below), whereas Glu_{487} helps to organize the necessary structure for catalysis by linking the NAD^+ and substrate-binding sites through ion-pairing interactions with two Arg (264 and 475) residues. Glu_{476} (not shown) provides a bridge between Arg_{475} and a structured water molecule lying near the amide portion of NAD^+ as seen in one published image of the active site. Other residues that do not H-bond to the cofactor in this conformation, but are conserved within its binding pocket, include Asn_{169} (shown) and Gly_{245}, Gln_{349} and Phe_{401} (not shown).

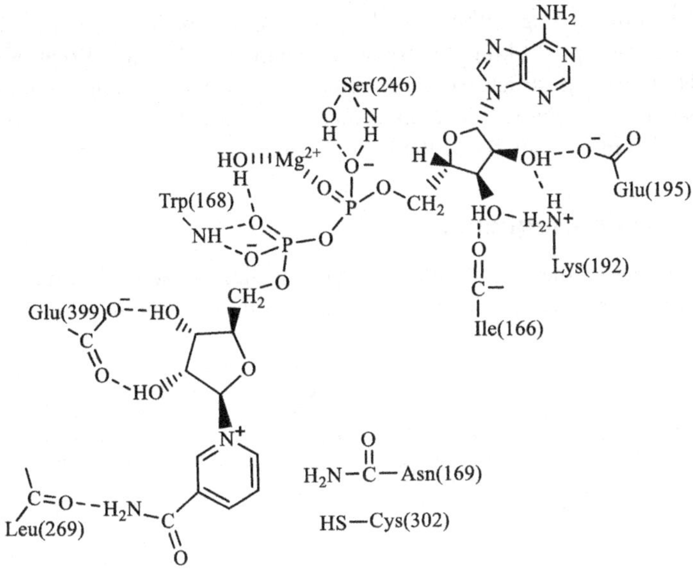

FIGURE 5.3 "Hydride transfer" binding pocket for NAD^+ in human ALDH2. *Adapted with permission from Perez-Miller, S. J. and Hurley, T. D.* Biochemistry ***42****, 7100–7109 (2003). Copyright 2003 American Chemical Society.*

After hydride transfer to the NAD^+, the reduced cofactor rotates slightly into the "hydrolysis" conformation, primarily involving motion of the nicotinamide portion of the molecule. That half of the cofactor in this conformation is shown in detail in Fig. 5.4. Several of the residues noted earlier are

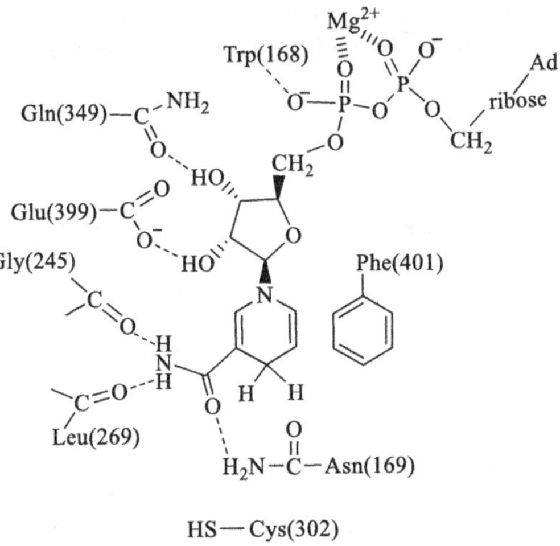

FIGURE 5.4 "Hydrolysis" binding pocket for NADH in human ALDH2. *Copyright 2003 American Chemical Society. See Fig. 5.3.*

now found interacting with the NADH by hydrogen bonding (Asn_{169}, Gly_{245}, Gln_{349}) or by a pi stacking, hydrophobic interaction (Phe_{401}). Rotation of the NADH affords the structured water molecule access to the intermediate thiohemiacetal (see below). The mechanistically key Glu_{268} lies astride the Glu_{476}, also in the vicinity of the NAD^+ amide.

5.4 Reaction sequence

An overview of the oxidation/reduction sequence is shown in Fig. 5.5. The first step (a) involves nucleophilic attack by a thiolate anion on the aldehyde

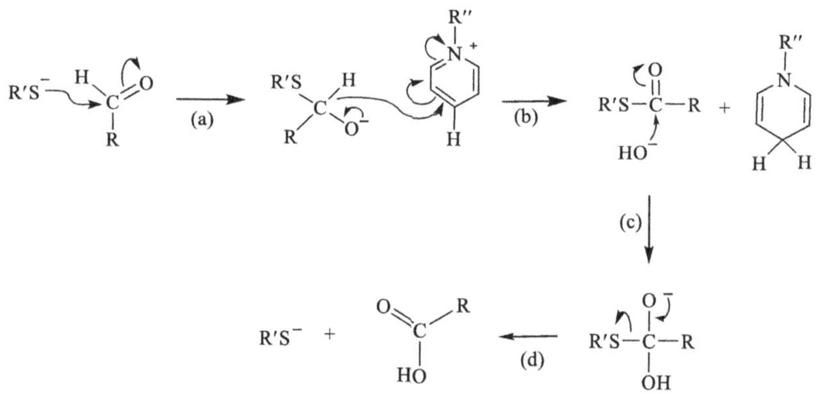

FIGURE 5.5 Overview of the chemistry catalyzed by ALDH2.

substrate to form a tetrahedral thiohemiacetal oxyanion intermediate. This anion then transfers a hydride ion to the pyridinium portion of NAD^+ to form an intermediate thioester and the reduction product (NADH) (step (b)). Hydrolysis of the thioester (steps (c) and (d)) restores the thiolate species and forms the substrate carboxylic acid.

5.5 Detailed mechanism and the role of active site residues

This is shown in Fig. 5.6. The reaction starts with the incorporation of NAD^+ into its ALDH2 binding pocket (step (a)), followed by binding of the substrate aldehyde to form a ternary complex (step (b)). There is general agreement that Glu_{268} activates the Cys_{302} for its nucleophilic attack on the carbonyl carbon. It is not clear whether it does this by the glutamate anion acting as a base (shown as step (c)) or as Glu-CO_2H helping to reduce the pKa of the thiol group. The tetrahedral intermediate formed in step (d) lies in an oxyanion hole, which includes the side chain amide nitrogen of Asn_{169} and the peptide nitrogen of Cys_{302}. Hydride transfer to the NAD^+ occurs in step (e). The structured water molecule within the active site is then

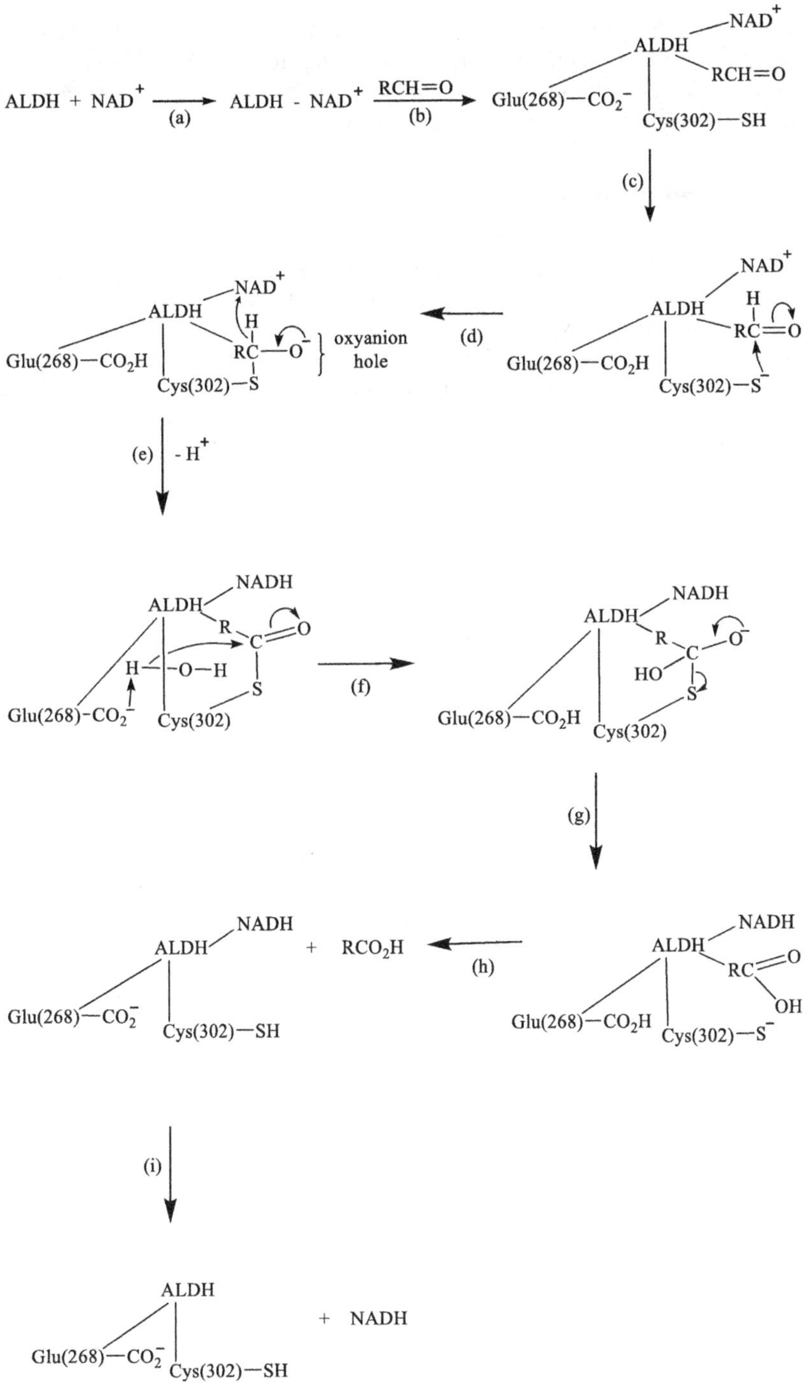

FIGURE 5.6 Mechanism and the role of active site residues in ALDH.

deprotonated by Glu_{268} and hydrolyzes the enzyme-bound thioester (steps (f) and (g); deacylation is the rate-determining step for ALDH2). Upon completion of hydrolysis, the reaction is completed by the sequential dissociation of the oxidized substrate (step (h)) and the NADH (step (i)).

Leading references

Chen, Y.; Zhu, J.-Y.; Hong, K. H.; Mikles, D. C.; Georg, G. I.; Goldstein, A. S.; Amory, J. K.; Schönbrunn, E. *ACS Chem. Biol.* **13**, 582–590 (2018); Buchman, C. D. and Hurley, T. D. *J. Med. Chem.* **60**, 2439–2455 (2017); Jung, K.; Hong, S.-H.; Ngo, H.-P.-T.; Ho, T.-H.; Ahn, Y.-J.; Oh, D.-K.; Kang, L.-W. *Int. J. Biol. Macromol.* **105**, 816–824 (2017); Moretti, A.; Li, J.; Donini, S.; Sobol, R. W.; Rizzi, M.; Garavaglia, S. *Sci. Rep.* **6**, 35710–35721 (2016); González-Segura, L.; Ho, K.-K.; Perez-Miller, S.; Weiner, H.; Hurley, T. D. *Chem. Biol. Interact.* **202**, 32–40 (2013); Marchitti, S. A.; Brocker, C.; Stagos, D.; Vasiliou, V. *Expert Opin. Drug Metab. Toxicol.* **4**, 697–720 (2008); Perez-Miller, S. J. and Hurley, T. D. *Biochemistry* **42**, 7100–7109 (2003).

Chapter 6

Arginase I

6.1 Arginase

Arginase (EC 3.5.3.1; human arginase I; HA1) is a hydrolytic, manganese (Mn^{2+}) metalloenzyme that exists as a pair of isozymes, arginase I, and arginase II. Arginase I hydrolyzes L-arginine into L-ornithine and urea. The enzyme is homotrimeric with an approximate molecular weight of 105 kDa. Each protomer contains a binuclear manganese cluster, with the manganese ions serving to coordinate, and ultimately, help in the deprotonation of a nucleophilic water molecule. The overall chemistry is shown in Fig. 6.1.

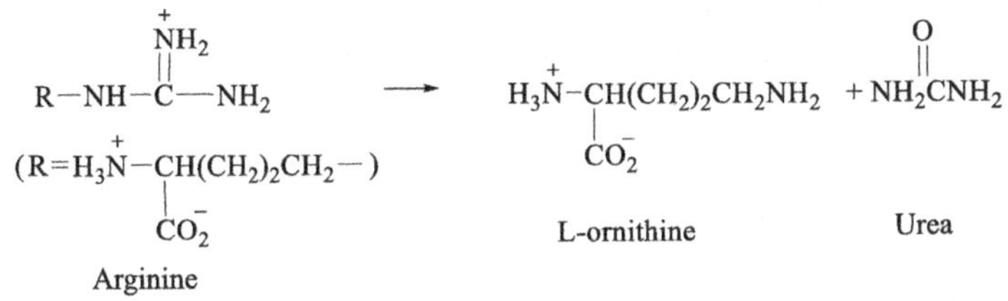

FIGURE 6.1 The overall chemistry catalyzed by arginase.

6.2 Physiological function

Arginase I is predominantly located in the cytoplasm of the liver and catalyzes the final step of the urea cycle. This metabolic sequence is a five-step process that ultimately leads to the elimination of ammonia in mammals. It has drawn interest as a target for cancer chemotherapy because it is upregulated in several cancer cell lines. Alternatively, the enzyme can be used to reduce arginine levels in cancer cell lines that require extracellular L-Arg.

6.3 Key structural features

An X-ray structure of the active site of human arginase I has been obtained for the enzyme bound to the inhibitor, 2(*S*)-amino-6-boronohexanoic acid (see Fig. 6.2). The boronate anion represents an excellent model for the transition state involved in the arginine hydrolysis (see below). Note, in particular, that His_{141} exists in its protonated state and can function as a source of protonation of

Enzyme Active Sites and their Reaction Mechanisms. DOI: https://doi.org/10.1016/B978-0-12-821067-3.00006-4

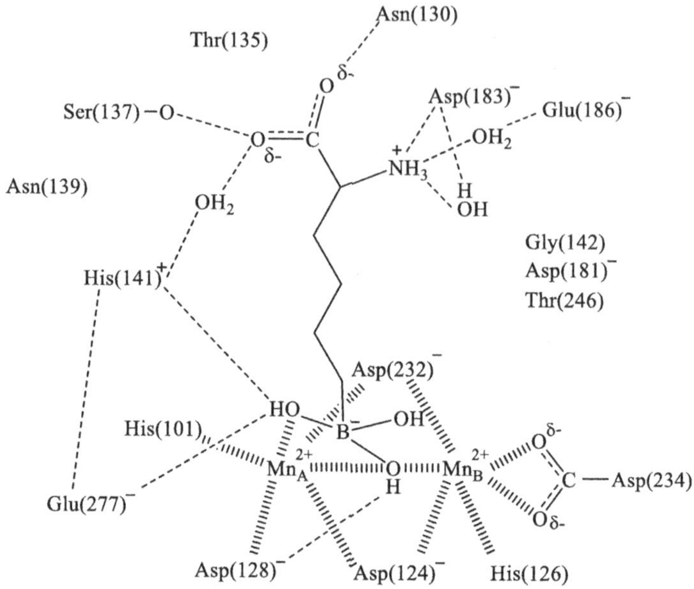

FIGURE 6.2 The active site of human arginase I containing 2(*S*)-amino-6-boronohexanoic acid bound as a boronate anion. *Adapted with permission from D'Antonio. E. L. and Christianson, D. W.* Biochemistry *50, 8018–8027 (2011). Copyright 2011 American Chemical Society.*

the ornithine product (see below). Several amino acids present within the active site, but not directly involved in the mechanism, are included in the figure.

6.4 Reaction sequence

The reaction involves hydrolysis of arginine's guanidino functional group into the amino acid, L-ornithine, and urea, the ultimate metabolite of the urea cycle. As is typical for a hydrolytic reaction involving an sp^2 hybridized center, a tetrahedral intermediate is key to the transformation (see Fig. 6.3).

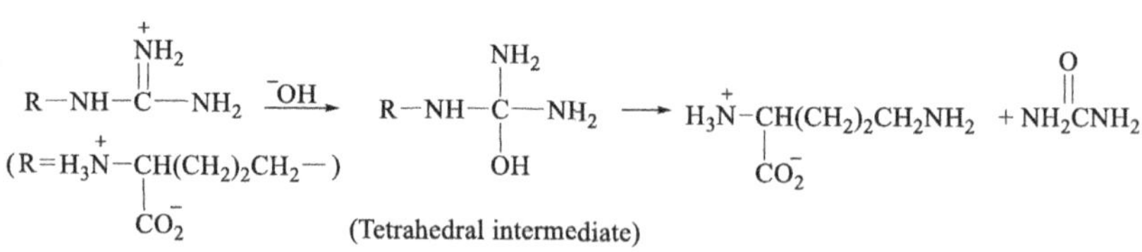

FIGURE 6.3 General mechanism for base-catalyzed hydrolysis of arginine.

6.5 Detailed mechanism and the role of the active site residues

The multistep hydrolytic reaction is shown in Fig. 6.4. The arginine is bound to the enzyme in step (a). AAs that interact with the metal ions,

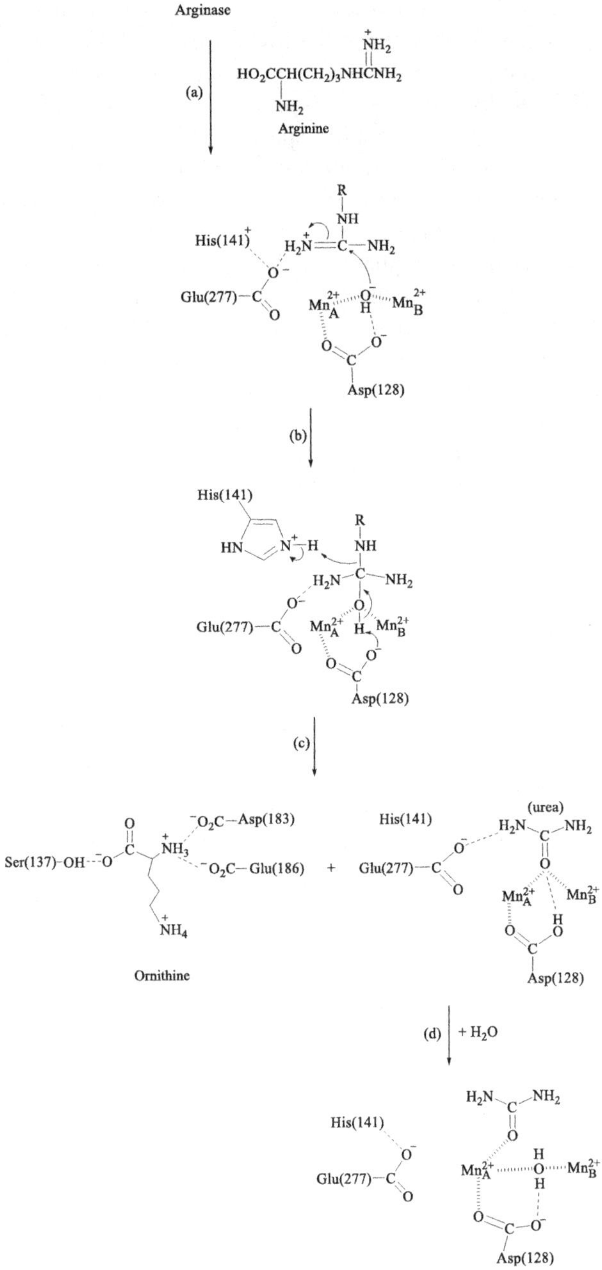

FIGURE 6.4 Detailed mechanism for the arginase-catalyzed hydrolysis of the guanidinium group of arginine showing the involvement of key amino acids at the active site. *Adapted from Costanzo, L. D. et al.* Proc. Natl. Acad. Sci. ***102****, 13058–13063 (2005); Copyright 2005 National Academy of Sciences, U.S.A.*

His_{101}, His_{126}, Asp_{124}, Asp_{232}, and Asp_{234}, have been omitted for clarity. Note that coordination of water to the Mn^{2+} ions (and Asp_{128}) has led to its deprotonation; coordination of water to metal ions increases the acidity of the O—H bond by as much as 9 pKa units relative to unbound water. This creates a strongly nucleophilic species which, in step (b), attacks the guanidinium group with the formation of a tetrahedral intermediate. Step (c) shows the collapse of this intermediate to the hydrolysis products, ornithine and urea. The Asp_{126} serves as a base in this reaction and His_{141} as an acid. Rotation of the His_{141} is required for this to occur. The ornithine structure in this figure shows AA residues observed interacting with the ornithine in enzyme crystals within which it has been embedded. The urea remains bridged to the two Mn^{2+} ions. Step (d) shows the reentry of water into the active site. Note that a second rotation of His_{141} restores its interaction with Glu(277). A proton transfer from Asp(128) to the ornithine returns this residue to its anionic state. Ultimately, the urea dissociates from the enzyme and the bridging water molecule is deprotonated to give a metal-bridged hydroxide ion (see "a" above).

Leading references

D'Antonio, E. L. and Christianson, D. W. *Biochemistry* **50**, 8018–8027 (2011); Stone, E. M.; Chantranupong, L.; Georiou, G. *Biochemistry* **49**, 10582–10588 (2010); Leopoldini, M.; Russo, N.; Toscano, M. *Chem. Eur. J.* **15**, 8026–8036 (2009); Costanzo, L. D.; Sabio, G.; Mora, A.; Rodriguez, P. C.; Ochoa, A. C.; Centeno, F.; Christianson, D. W. *Proc. Natl. Acad. Sci.* **102**, 13058–13063 (2005); Christianson, D. W. *Acc. Chem. Res.* **38**, 191–201 (2005); Cama, E.; Elleluori, D. M.; Emig, F. A.; Shin, H.; Kim, S. W.; Kim, N. N.; Traish, A. M.; Ash, D. E.; Christianson, D. W. *Biochemistry* **42**, 8445–8451 (2003); Cox, J. D.; Cama, E.; Colleluori, D. M.; Pethe, S.; Boucher, J.-L.; Mansuy, D.; Ash, D. E.; Christianson, D. W. *Biochemistry* **40**, 2689–2701 (2001).

Chapter 7

Carbonic anhydrase II

7.1 Human carbonic anhydrase II

Human carbonic anhydrase II (EC 4.2.1.1; HCA II; α-CAII; HCA2; carbonate dehydratase) is a member of a superfamily of metalloenzymes, consisting of six known subfamilies (α, β, γ, δ, ζ, η), that is ubiquitous among living organisms. All reversibly interconvert carbon dioxide and water to bicarbonate and H^+. The only one of these subfamilies expressed by vertebrates is the α class, within which 15 isoforms (I–XV) have been identified. The most extensively studied of these is the human intracellular, cytosolic enzyme, α-CAII (HCA II). The chemical reactions catalyzed by the CAs are shown in Fig. 7.1.

$$CO_2 + H_2O \rightleftharpoons HCO_3^- + H^+$$

FIGURE 7.1 The reversible chemical transformations catalyzed by carbonic anhydrase.

7.2 Physiological function

HCA II is found in many different organs and cell types; it is the major nonhemoglobin protein in human red cells. The enzyme maintains the acid–base balance in blood and other tissues and serves to transport CO_2 out of tissues. It is known to play a role in numerous physiological processes, including respiration, calcification, tumorigenesis, pH regulation, and gluconeo-, urea- and lipogenesis. It has been associated with numerous diseases, such as cancer, glaucoma, edema, epilepsy, and morbid obesity, and as such, has been the subject of numerous inhibition studies for drug development.

7.3 Key structural features

The active site of HCA II lies at the bottom of a cone-shaped cavity. All members of the α-class utilize divalent Zn to catalyze the chemistry. The metal and its ligands lie adjacent to a hydrophilic region that facilitates proton and bicarbonate transport. The key residues here are His_{64}, Leu_{198}, Thr_{198}, and Thr_{199}. The Thr_{199} has been referred to as a gatekeeper residue,

Enzyme Active Sites and their Reaction Mechanisms. DOI: https://doi.org/10.1016/B978-0-12-821067-3.00007-6

being thought to control the potential access of substrates to the Zn ion, and to possibly assist in the release of the bicarbonate product. It also plays a role in orienting the CO_2 relative to the Zn-bound oxygen (see below). All of these functions are influenced by its role as a H-bond donor to Glu_{106}. His_{64} will be seen to play a critical role in proton transport (e.g., the "proton shuttle"). A second, hydrophobic, region of the active site lies opposite and it is here that the CO_2 substrate is bound. Key residues are Trp_{209}, Val_{207}, Val_{143}, and Val_{121}. The zinc is attached to three His residues (His_{94}, His_{96}, His_{119}) and a water molecule in a tetrahedral configuration. The Zn-bound residues are stabilized by hydrogen-bonding to nearby ("second-level"; "indirect") AAs-His_{94} to Gln_{92}, His_{96} to Asn_{244}, and His_{119} to Glu_{117}. The active site containing the carbon dioxide substrate is schematically depicted in Fig. 7.2. Note that the active sites for the holoenzyme (with zinc), the holoenzyme with bound CO_2, and the holoenzyme with bound bicarbonate have all been found to be little changed from the structure found for the apoenzyme. One notable difference is that a water molecule, seen in the holo-structure and termed the "deep water", lies approximately in the CO_2 binding pocket and is displaced by the substrate upon its entry there. There are also several other water molecules within the active site (not shown).

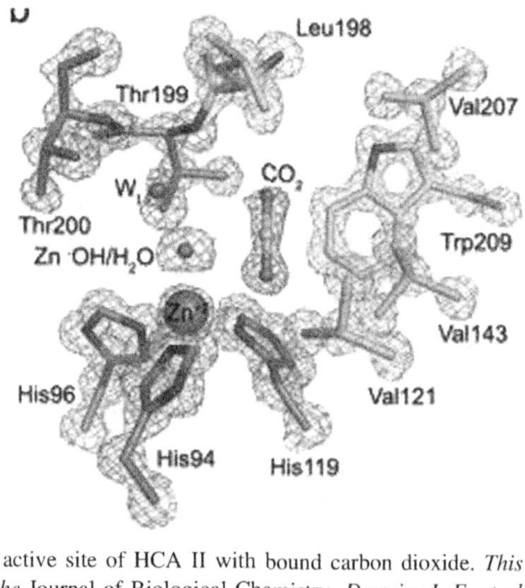

FIGURE 7.2 The active site of HCA II with bound carbon dioxide. *This research was originally published in the* Journal of Biological Chemistry. *Domsic, J. F. et al. Entrapment of carbon dioxide in the active site of carbonic anhydrase II.* J. Biol. Chem. *2008; 283; 30766–30771; The American Society for Biochemistry and Molecular Biology.*

7.4 Reaction sequence

The reaction catalyzed by HCA II (see Fig. 7.1) is deceptively simple-looking. However, to drive this reaction to the right requires that a nucleophilic hydroxide species be generated at physiological pH. The key, of course, is the coordination of the oxygen to the zinc, which lowers its pKa by 7 orders of magnitude. The chemistry is considered to proceed by a two-step, "ping-pong" mechanism, as shown in Fig. 7.3. The "B" in the second step is a buffer molecule lying within the bulk solvent.

$$EZnOH^- + CO_2 + H_2O \rightleftharpoons EZnOH_2 + HCO_3^-$$

$$EZnOH_2 + B \rightleftharpoons EZnOH^- + BH^+$$

FIGURE 7.3 The "ping-pong" mechanism of HCA II. *HCA II*, Human carbonic anhydrase II.

7.5 Detailed mechanism and the role of active site residues

A summary mechanism is shown in Fig. 7.4. In step (a), His_{64} facilitates the deprotonation of a proton from zinc-bound water through the intermediacy of bridging water molecules. Neither the exact number of water molecules nor the details of this "proton shuttle" pathway are known with certainty, but two intermediate waters have been calculated to be optimal. There are two conformations of His_{64} that are seen in X-ray structures—one pointing inward toward the Zn, and one pointing away toward the bulk solvent. It is generally accepted that the "inward" conformation is dominant in this proton abstraction step, with a shift to the "outward" conformation for delivery of the proton to the bulk medium (see step (d) below). As noted earlier, the pKa of the zinc-bound water is lowered by its attachment to the metal ion.

The addition of CO_2 leads to its encapsulation in the hydrophobic pocket. The Thr_{199} helps to stabilize the substrate and orients it for nucleophilic attack by the zinc-bound hydroxide ion as shown in step (b). This leads to a bound bicarbonate anion, which reorients in step (c) to a Zn bidentate-bound complex. Note that earlier proposals do not include such a species but there are X-ray images that support its presence. The reaction concludes in step (d) wherein the "proton shuttle" delivers a proton to a buffer molecule in the bulk medium, and a water molecule replaces the product bicarbonate ion. This final proton transfer is known to be the reaction's rate-determining step. All steps in this mechanism are reversible.

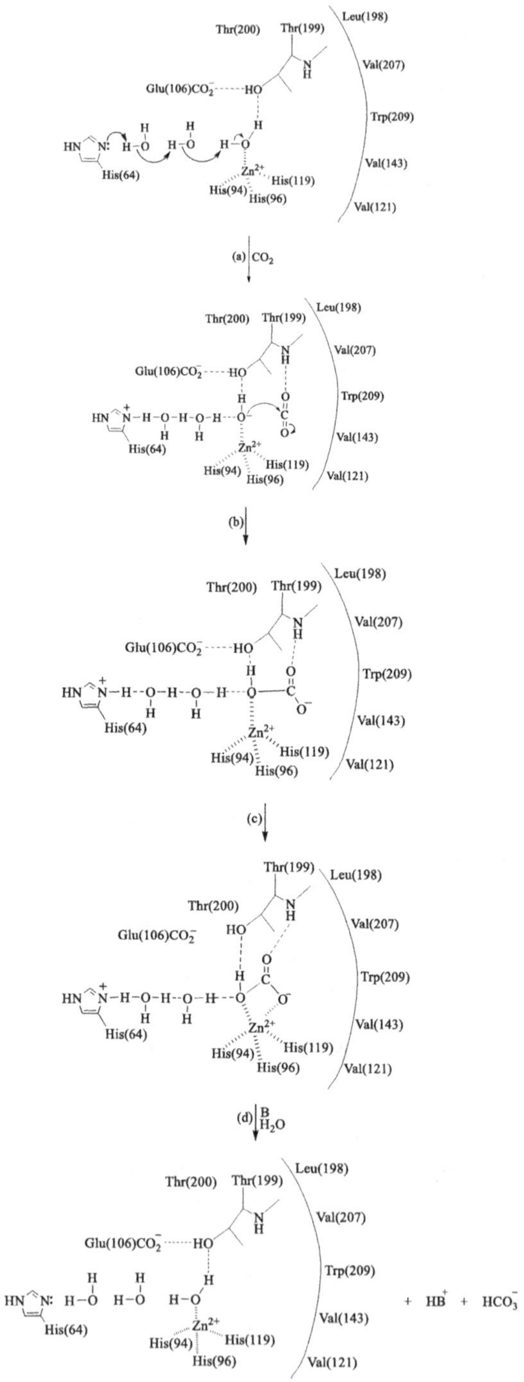

FIGURE 7.4 Detailed mechanism for the hydration of carbon dioxide catalyzed by HCA II. *HCA II*, Human carbonic anhydrase II.

Leading references

Supuram, C. T. *Biochem. J.* **473**, 2023–2032 (2016); Aggarwal, M.; Kondeti, B.; Tu, C.; Maupin, C. M.; Silverman, D. N.; McKenna, R. *IUCrJ* **1**, 129–135 (2014); Mikulski, R. M. and Silverman, D. N. *Biochim. Biophys. Acta* **1804**, 422–426 (2010); Domsic, J. F. and McKenna, R. *Biochim. Biophys. Acta* **1804**, 326–331 (2010); Gilmour, K. M. *Comp. Biochem. Physiol., Part A* **157**, 193–197 (2010); Domsic, J. F.; Avvaru, B. S.; Kim, C. U.; Gruner, S. M.; Agbandje-McKenna, M. *J. Biol. Chem.* **283**, 30766–30771 (2008); Fisher, S. Z.; Maupin, C. M.; Budayova-Spano, M.; Govindasamy, L.; Tu, C.; Agbandje-McKenna, M.; Silverman, D. N.; Voth, G. A.; McKenna, R. *Biochemistry* **46**, 2930–2937 (2007).

Chapter 8

Carboxypeptidase A

8.1 Carboxypeptidase A

Carboxypeptidase A (EC 3.4.17.1; CPA) is a hydrolytic enzyme typically isolated from the bovine pancreas. It is a zinc metalloprotease, one of four major families of protease enzymes, (e.g., enzymes that catalyze the hydrolysis of peptide amide bonds). Other protease families include serine (see Chapter 9: Chymotrypsin), cysteine (see Chapter 18: Hepatitis C NS 2/3 Protease), and aspartic proteases (see Chapter 19: HIV-1 Protease). Zinc is the second most abundant metal in humans and is utilized in multiple roles. The overall reaction for the (base-catalyzed) hydrolysis of a peptide bond is shown in Fig. 8.1.

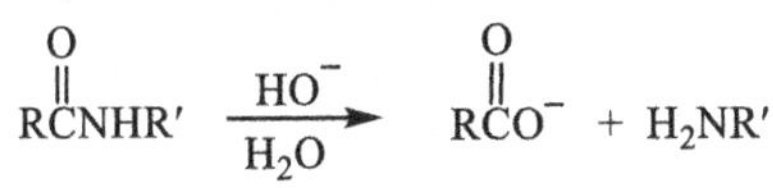

FIGURE 8.1 The base-catalyzed hydrolysis of a peptide bond.

8.2 Physiological function

Carboxypeptidase is formed in the pancreas from the zymogen, procarboxypeptidase. It is a digestive exopeptidase, an enzyme that cleaves a terminal residue from a polypeptide. CPA cleaves the C-terminal amino acid with a preference for AAs with a large, hydrophobic side chain.

8.3 Key structural features

Bovine CPA contains one Zn atom and consists of a single polypeptide chain of 307 amino acids. It normally exists as a noncovalent tetramer. Those residues found to be important to the active site are Glu_{270}, Arg_{71}, Arg_{127}, Asn_{144}, Arg_{145}, and Tyr_{248}. In addition, His_{196}, Glu_{72}, and His_{69} are bound to the zinc atom. A rendering of the active site with embedded *N*-benzoyl-alanylphenylalanine is shown in Fig. 8.2.

Enzyme Active Sites and their Reaction Mechanisms. DOI: https://doi.org/10.1016/B978-0-12-821067-3.00008-8

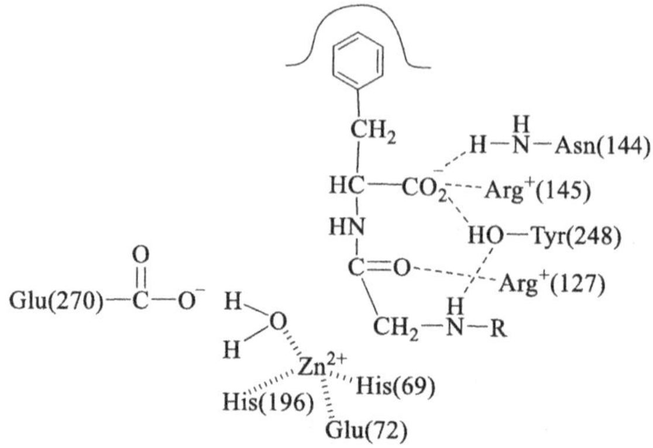

FIGURE 8.2 The carboxypeptidase A active site containing *N*-benzoyl-alanylphenylalanine. *Adapted with permission from Christianson, D. W. and Lipscomb, W. N.* Acc. Chem Res. *22, 62–69 (1989). Copyright 1989 American Chemical Society.*

8.4 Reaction sequence

Unlike the mechanism for, for example, chymotrypsin, the current view of CPA catalytic activity favors hydrolysis of the peptide linkage *without* the formation of a stable, acyl-enzyme intermediate. Rather, water itself nucleophilically attacks the amide through activation by coordination with the zinc atom and deprotonation by the glutamate (see below). This is depicted in the simple hydrolytic sequence of Fig. 8.3. (Note that there has been active debate about an alternative mechanism in which Glu_{270} does form an intermediate anhydride upon attack on the amide group).

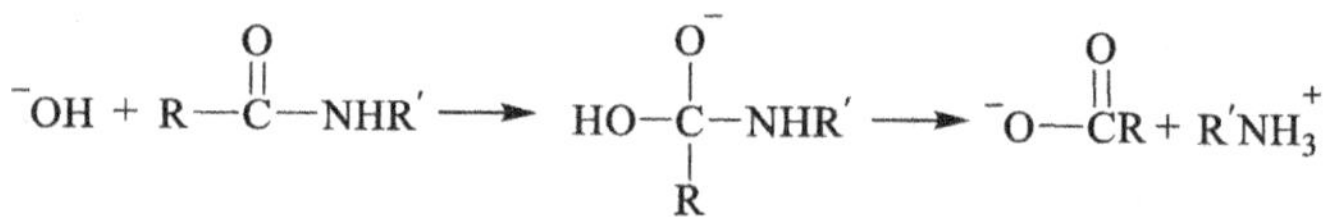

FIGURE 8.3 General mechanism for the base-catalyzed hydrolysis of a peptide bond.

8.5 Detailed mechanism and the role of the active site residues. The "promoted water" mechanism

A detailed mechanism is outlined in Fig. 8.4, where the peptide substrate is *N*-benzoyl-alanylphenylalanine. This mechanism involves activation of water as a nucleophile by the association with the zinc atom, deprotonation of this bound water by Glu (270), and nucleophilic attack by the water on the amide carbonyl group (see step (a)). Coordination of water to metal ions increases the acidity of the O—H bond by as much as 9 pKa units relative to unbound water. The zinc atom also plays a role in stabilizing the

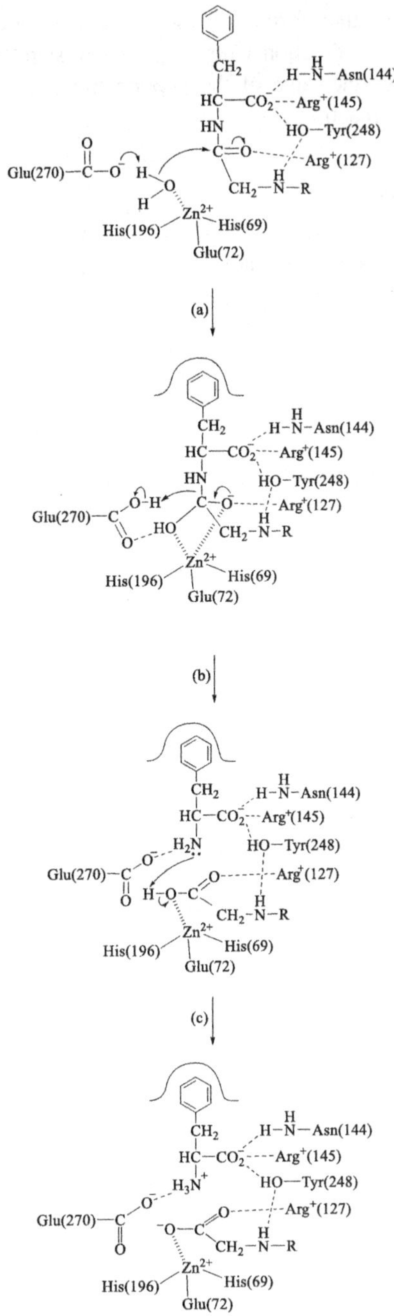

FIGURE 8.4 Detailed mechanism for the carboxypeptidase-catalyzed hydrolysis of a peptide bond showing the involvement of key amino acids at the active site. See Fig. 8.2. *Copyright 1989 American Chemical Society.*

tetrahedral intermediate thus formed. Attack on the peptide carbonyl oxygen is also facilitated by its coordination with Arg(127). In step (b), the zinc atom and the Glu (270) facilitate cleavage of the peptide linkage, and proton transfer in step (c) completes the reaction.

Leading references

Shushanyan, M.; Khoshtariya, D. E.; Tretyakova, T.; Makharadze, M.; van Eldik, R. *Biopolymers* **95**, 852–868 (2011); Christianson, D. W. and Lipscomb, W. N. *Acc. Chem Res.* **22**, 62–69 (1989).

Chapter 9

Chymotrypsin

9.1 α-Chymotrypsin

α-Chymotrypsin (EC 3.4.21.1; chymotrypsinogen A) is a "hydrolytic enzyme" member of the super-family of serine proteases, enzymes that hydrolytically cleave peptide bonds utilizing a serine hydroxyl group as a nucleophile at the active site. The most extensively studied is bovine pancreatic chymotrypsin. Other enzymes within this classification include elastase, trypsin, thrombin, choline esterase, and subtilisin (a bacterial protease). The overall reaction is shown in Fig. 9.1. Of special note is that chymotrypsin

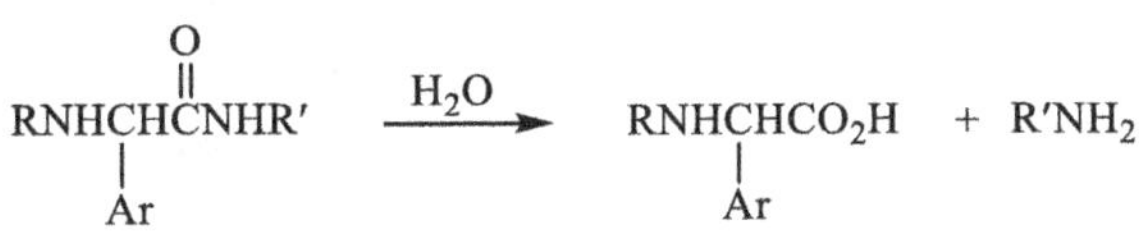

FIGURE 9.1 The overall chemistry catalyzed by α-chymotrypsin.

cleaves peptide bonds that are on the C-terminal side of an amide linkage, which contains an aromatic (Tyr, Phe, and Trp) side chain.

9.2 Physiological function

This enzyme is a digestive peptidase the purpose of which is to catalyze the hydrolysis of the peptide bonds of protein foods in the mammalian gut. α-Chymotrypsin is formed from chymotrypsinogen, a "zymogen" consisting of 245 amino acid residues. Zymogens are inactive enzyme precursors that are processed to an active enzyme by cleavage at one or more peptide bonds. Chymotrypsinogen is synthesized in the pancreas and secreted into the small intestine where it has a relatively short lifetime and is converted into α-chymotrypsin.

9.3 Key structural features

The residues critical to the hydrolytic mechanism are His_{57}, Asp_{102}, and Ser_{195}, often referred to as a "catalytic triad." The specificity for a neighboring aryl side chain derives from a "hydrophobic pocket" within the active

Enzyme Active Sites and their Reaction Mechanisms. DOI: https://doi.org/10.1016/B978-0-12-821067-3.00009-X

site. An oxide anion is formed as an intermediate during the hydrolysis (see below); its stabilization (and that of the transition state leading to it) by an "oxyanion hole" is instrumental in the catalytic sequence (Fig. 9.2).

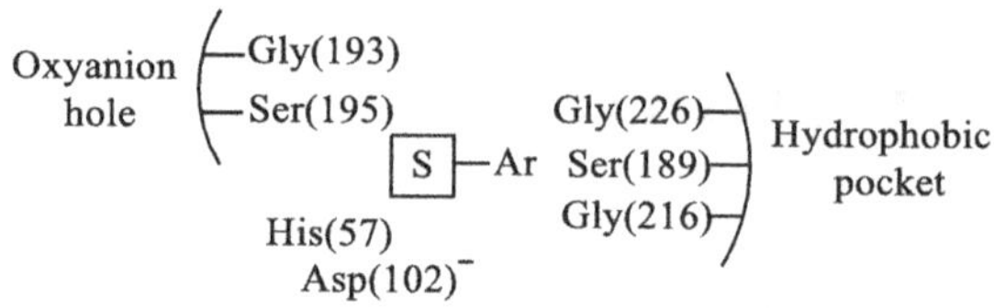

FIGURE 9.2 Schematic rendering of the active site for α-chymotrypsin showing key residues (S = substrate).

9.4 Reaction sequence

The hydrolytic reaction involves the nucleophilic addition of the α-chymotrypsin Ser_{195}-OH to an amide linkage of the target protein to form a "tetrahedral intermediate" typical of the hydrolysis of carboxylic acid derivatives. This intermediate fragments to form an acyl-enzyme ester and the amine portion of the target peptide bond. Hydrolysis of the acyl enzyme then releases the carboxylic acid portion of the peptide bond and regenerates the α-chymotrypsin Ser_{195}-OH. Formation of the acyl-enzyme ester is the "rate-determining" step. The sequence is outlined in Fig. 9.3.

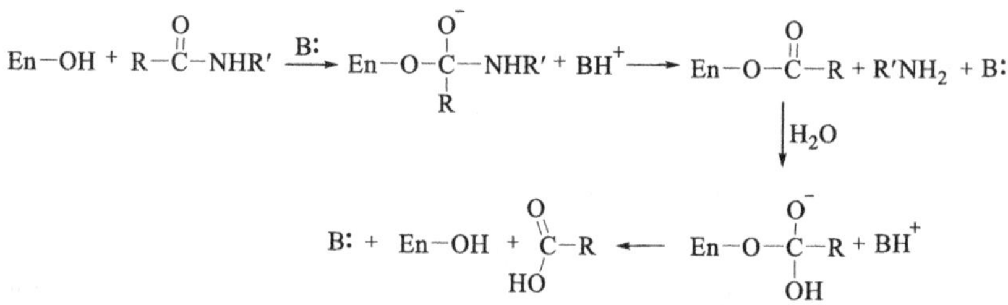

FIGURE 9.3 General reaction sequence for chymotrypsin-catalyzed hydrolysis of a peptide bond.

9.5 Detailed mechanism and the role of the active site residues

A more detailed description of the mechanism is shown in Fig. 9.4. The mechanism shown here is characteristic of serine proteases (see above). In step (a) His_{57} deprotonates the Ser_{195} hydroxyl group to enable nucleophilic attack on the amide carbonyl group. The ability of the less basic histidine N3 to deprotonate the more basic alcohol group of Ser_{195} has been a conundrum and is seemingly contra-thermodynamic. One explanation is that the third

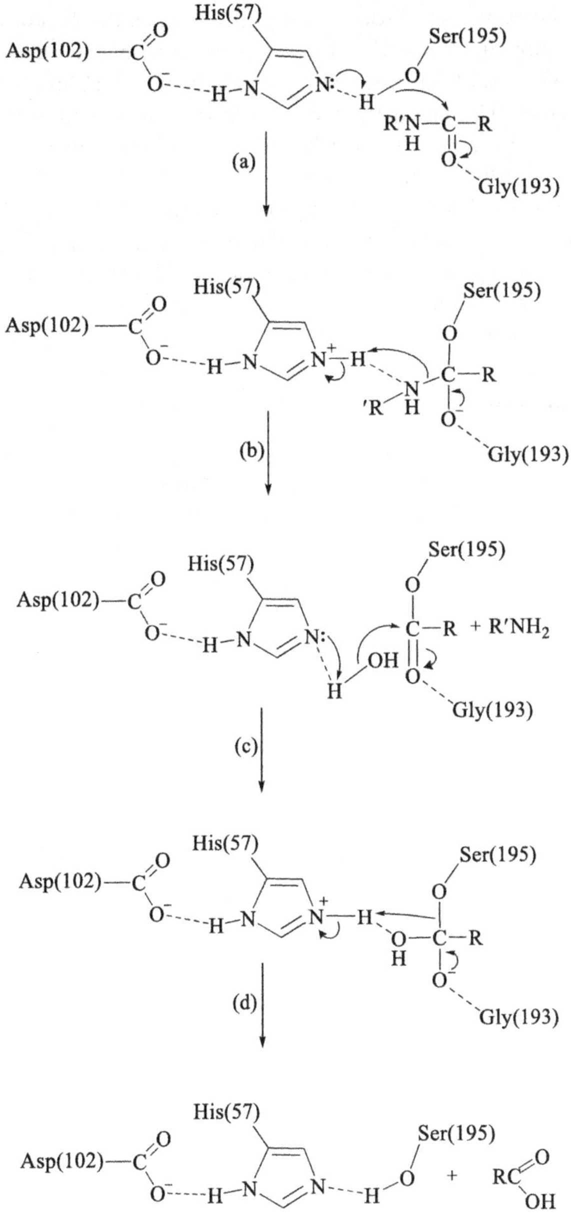

FIGURE 9.4 Detailed mechanism for chymotrypsin-catalyzed hydrolysis showing the involvement of key amino acids at the active site.

member of the active site "triad," Asp_{102}, facilitates the deprotonation reaction by forming an unusually strong, short, "low-barrier hydrogen bond" with His_{57}. This has been disputed, and a network of H-bonds is suggested as an alternative. The anion formed by step (a) moves within the active site toward an adjacent region termed the "oxyanion hole" where it is stabilized by hydrogen bonding with main-chain N—H bonds from Gly_{193} (and the Ser_{195}; not shown). The amide linkage cleaves in step (b) with the release of an amine and the formation of an acylated enzyme. In step (c) His_{57} again acts as a base in deprotonating a water molecule that attacks the serine ester. Cleavage in step (d) releases the Ser_{195} and the carboxylic acid portion of the original peptide linkage.

Leading references

Fuhrmann, C. N.; Daugherty, M. D.; Agard, D. A. *J. Am. Chem. Soc.* **128**, 9086–9102 (2006); Frey, P. A. *J. Phys. Org. Chem.* **17**, 511–520 (2004); Kollman, P. A.; Kuhn, B.; Donini, O.; Perakyla, M.; Stanton, R.; Bakowies, D. *Acc. Chem. Res.* **34**, 72–79 (2001).

Chapter 10

Citrate synthase

10.1 Citrate synthase

Citrate synthase (EC 2.3.3.1, formerly 4.1.3.7; (*S*-)citrate synthase; CS) is the first of a series of eight enzymes involved in the citric acid cycle. The citric acid cycle is also known as the tricarboxylic acid cycle and the Krebs cycle. CS is found in virtually all living cells. The enzyme-catalyzed reaction occurs in two steps (see below) with the overall conversion shown in Fig. 10.1. Note that citric acid is "prochiral," and it is the pro-*S* acetic acid

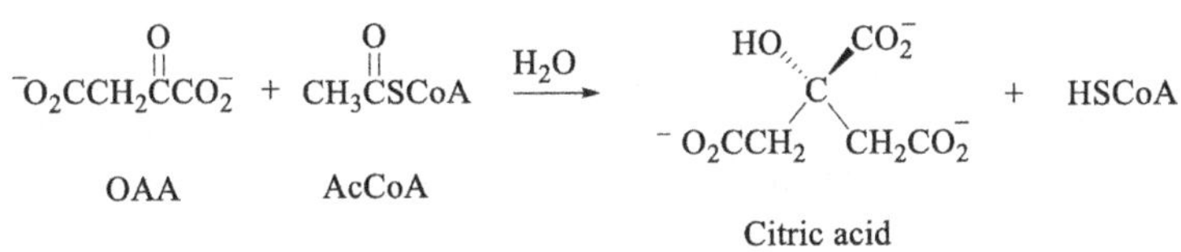

FIGURE 10.1 The overall chemistry catalyzed by citrate synthase.

side chain (right side in Fig. 10.1) that originates from the acetyl group of acetyl-CoA (AcCoA; see Chapter 15: Dihydrolipoyl Transacetylase for the structure of CoA). It is interesting that this is the only reaction of the Krebs cycle that generates a C—C bond.

10.2 Physiological function

The citric acid cycle is the key metabolic pathway responsible for the oxidative degradation of amino acids, fatty acids, and carbohydrates and a source of numerous biosynthetic intermediates. CS is synthesized in the cytosol and then transported into the mitochondria.

10.3 Key structural features

CS is typically isolated from pig or chicken heart muscle. It is a dimer of two identical subunits, with each of these binding one molecule each of the two reactants, oxaloacetic acid (OAA) and AcCoA. Each of the enzyme subunits has two domains—one large and one small, with the active site contained within a cleft between the two domains that lies 15 Å deep from the

Enzyme Active Sites and their Reaction Mechanisms. DOI: https://doi.org/10.1016/B978-0-12-821067-3.00010-6

enzyme surface. The active site utilizes residues from both domains and contains separate binding sites for OAA and AcCoA. The OAA binds first, and in so doing, causes a conformational rearrangement of the smaller domain that generates the binding site for AcCoA. Several conformations of the enzyme have been observed by X-ray analysis, each classified as either "closed" or "open." The substrates bind tightly, and the chemistry occurs, within the closed form. The substrates are only weakly bound in open forms and these are associated with the ultimate release of the products into the medium.

The AcCoA is compact within the active site (an elongated structure is more typical) and shows an internal H-bond from the hydroxyl group of the pantoic acid portion of the CoA side chain to N-7 of the adenosine ring. There is also evidence for a tightly bound water molecule that links the pantoic acid-derived amide carbonyl group to the ribose hydroxyl group. X-ray analysis of the enzyme containing malate as the substrate shows the AcCoA carbonyl group H-bonded to His274 and a water molecule, and the AcCoA methyl group in close proximity to Asp_{375} [the basic residue responsible for the abstraction of a methyl group proton (see below)]. This is seen also in the active site structure obtained with OAA and shown in Fig. 10.2. It should be noted that the structure in Fig. 10.2 (and thus the proposed mechanism) differs markedly from those published in 1990 using carboxymethyl CoA, and in 1991 using malate and AcCoA Karpusas et al., 1990, 1991). Those structures, and the mechanism derived from them, show an involvement of two histidine residues not present in Fig. 10.2.

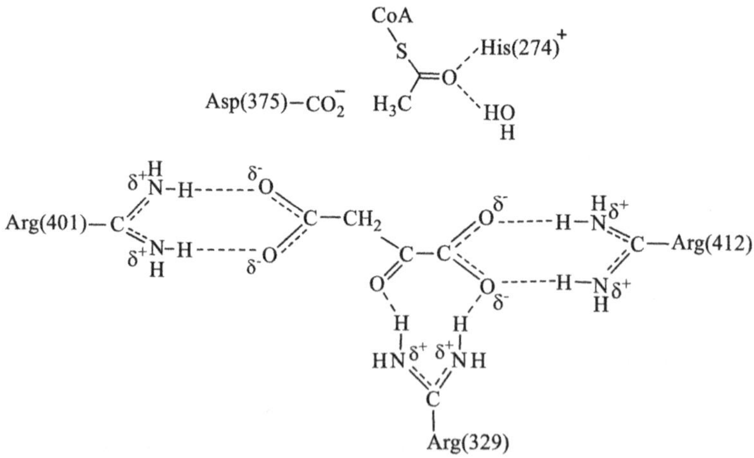

FIGURE 10.2 CS active site containing OAA and AcCoA. *AcCoA*, Acetyl-CoA; *CS*, citrate synthase; *OAA*, oxaloacetic acid. *Adapted with permission from van der Kamp et al.* J. Phys. Chem. B *114, 11303–11314 (2010); Copyright 2010 American Chemical Society.*

10.4 Reaction sequence

The enzyme creates a C—C bond through the condensation of acetyl-CoA and oxaloacetate (OAA). The resulting (*S*)-citryl-CoA is then hydrolyzed by the same enzyme to form citrate and CoA. The overall conversion involves a mixed aldol/Claisen condensation in which a C—C bond is generated between the α-methyl carbon of acetyl-CoA and the keto group of OAA. The resulting hydroxyl thioester is then hydrolyzed to citrate and HSCOA. The reaction sequence is shown in Fig. 10.3.

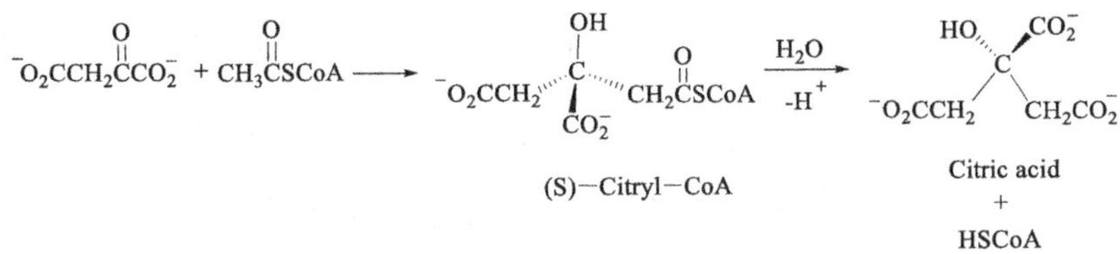

FIGURE 10.3 Generalized reaction sequence for the condensation of OAA with acetyl-CoA. *OAA*, Oxaloacetic acid.

10.5 Detailed mechanism and the role of active site residues

There are several challenges for the enzyme in accomplishing this chemistry. First, the enolization requires a relatively unacidic hydrogen to be abstracted by a weak base. Second, the enolate anion thus formed needs to be stabilized. Third, the carbonyl group of the OAA must be activated sufficiently for attack by an anion even while surrounded by the anionic carboxylate groups. Three AA residues, His_{274}, Asp_{375}, and Arg_{329} have been identified as key to the success of the enzyme in solving these challenges.

A detailed mechanism is presented in Fig. 10.4. Step (a) involves the deprotonation of the methyl group of acetyl-CoA by Asp_{375}, with His_{274} and the active site water polarizing the thioester carbonyl group, and stabilizing the resultant enolate anion (earlier literature favored the formation of the neutral enol). In the second (condensation) step (b), the enolate anion attacks the carbonyl group of OAA, which is activated by a salt bridge with, and H-bonding to, Arg_{329}. Several noteworthy points about this chemistry: (1) the earlier literature invokes His_{320} for the role played by the arginine residue, (2) computational evidence has been presented for some enhancement of both steps by His_{238}, which is adjacent to OAA in the active site, (3) elegant studies, utilizing a methyl group in AcCoA that is made up of the three hydrogen isotopes, have shown that step (a) and step (b) occur on opposite sides of the tetragonal carbon, that is, with inversion, (4) the active site orients the OAA with the *si* face of the carbonyl group facing the AcCoA enolate, so that the new C—C bond is formed stereospecifically on this side. The details of the hydrolysis step (c) are not yet known.

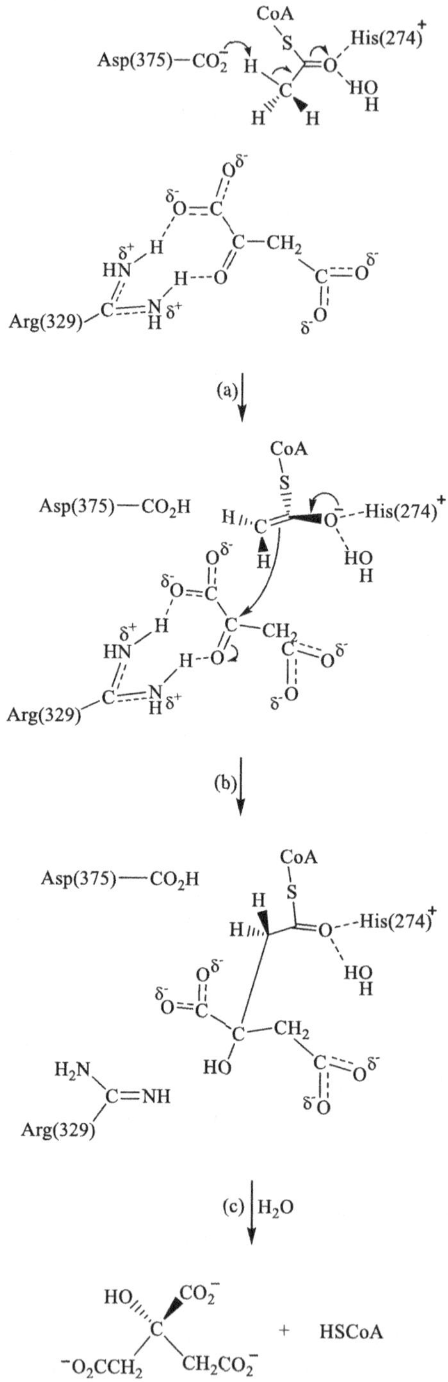

FIGURE 10.4 Detailed mechanism for the citrate synthase-catalyzed condensation of OAA and acetyl-CoA showing the involvement of key amino acids at the active site. See Fig. 10.2. *OAA*, Oxaloacetic acid. *Copyright 2010 American Chemical Society.*

Leading references

Bello, D.; Rubanu, M. G.; Bandaranayaka, N.; Götze, J. P.; Bühl, M.; O'Hagan, D. *ChemBioChem* **20**, 1174–1182 (2019); Bennie, S. J.; van der Kamp, M. W.; Pennifold, R. C. R.; Stella, M.; Manby, F. R.; Mulholland, A. J. *J. Chem. Theory Comput.* **12**, 2689–2697 (2016); Aleksandrov, A.; Zvereva, E.; Field, M. *J. Phys. Chem. B* **8**, 4505–4513 (2014); van der Kamp, M. W.; Żurek, J.; Manby, F. R.; Harvey, J. N.; Mulholland, A. J. *J. Phys. Chem. B* **114**, 11303–11314 (2010); Kurz, L. C.; Constantine, C. Z.; Jiang, H.; Kappock, T. J. *Biochemistry* **48**, 7878–7891 (2009); Remington, S. J. *Curr. Opin. Struct. Biol.* **2**, 730–735 (1992); Eggerer, H.; Buckel, W.; Lenz, H.; Wunderwald, P.; Cornforth, J. W.; Donninger, C.; Mallaby, R.; Redmond, J. W.; Gottschalk, G. *Nature* **20**, 517–519 (1970); Rétey, J.; Lüthy, J.; Arigoni, D. *Nature* **226**, 519–521 (1970).

Chapter 11

Cytochrome P450cam

11.1 Cytochrome P450cam

Cytochrome P450cam (EC 1.14.15.1; camphor 5-monooxygenase; CYP101A1) is a member of the ubiquitous cytochromes P450 family (CYPs) that are found in all forms of life. P450cam is a bacterial (from *Pseudomonas putida*), single-chain polypeptide that utilizes a Fe_2S_2 reductant (putidaredoxin; Pdx) to hydroxylate camphor. It is a monooxygenase (using only one oxygen from the O_2 molecule; see also peptidylglycine α-hydroxylating monooxygenase), "Type 1," heme-thiolate enzyme. The overall transformation, in which an unactivated C-H bond is hydroxylated, is shown below in Fig. 11.1.

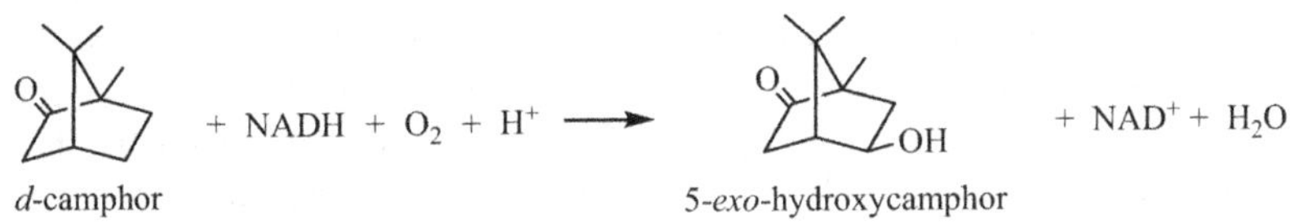

FIGURE 11.1 Overall reaction of P450cam.

Because of the extensive research on this particular enzyme, including the determination of the structure of its active site by X-ray analysis, it has been considered a prototypical example of this diverse family. The role of the iron in the hydroxylation chemistry has analogies with the mechanism of cleavage of L-Trp by IDO1.

11.2 Physiological function

CYPs detoxify exogenous chemicals by converting typically lipophilic xenobiotics into more hydrophilic, and therefore more readily metabolized, products. They are also important in the biosynthesis and metabolism of, for example, steroid hormones and fatty acids. In general, hydroxylation of substrates by CYPs is highly regio- and stereospecific.

11.3 Key structural features

The central heme structure is shown as Fig. 11.1 in the "IDO1" chapter. At the outset, a molecule of water and a thiolate residue (Cys_{357}) are *trans* axially bound

Enzyme Active Sites and their Reaction Mechanisms. DOI: https://doi.org/10.1016/B978-0-12-821067-3.00011-8

to the Fe^{III} atom. Fig. 11.2 shows several of the critical residues present at the camphor-bound active site. Hydrogen bonding of Cys_{357} with Phe_{351} and Gly_{359}

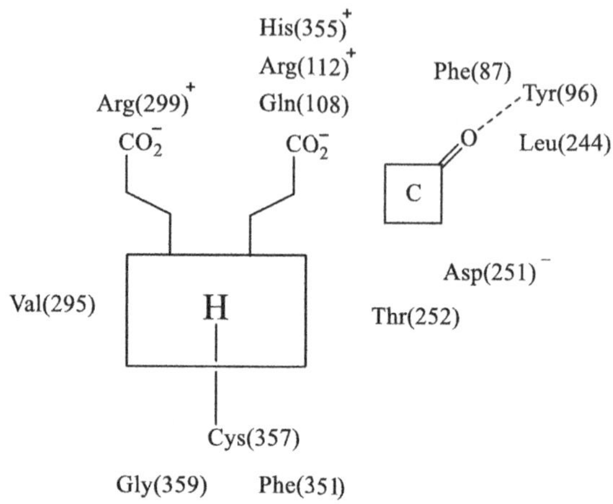

FIGURE 11.2 Schematic rendering of the active site with bound camphor showing key residues. "H" is the heme molecule and "C" the camphor substrate.

plays an important role in adjusting the redox potential of the heme iron, since the more negative charge residing on the thiolate, the less reducible is the ferric atom. His_{355}, Gln_{108}, and Arg_{112} form hydrogen bonds with the heme 6-propionate, and Arg_{299} with the 7-propionate. Upon binding of camphor into the active site, the distal, axial water molecule is displaced, as are several other nearby water molecules. Tyr_{96} forms a hydrogen bond to the camphor carbonyl group, which also interacts with Leu_{244} and Phe_{87}. Val_{295} interacts with the heme methyl groups. The camphor molecule is oriented so as to bring the C-5 carbon closest to the heme iron. The enzyme exists in an "open" form without the substrate and a "closed" form when the substrate is present. Arg_{112}, Thr_{252}, and Asp_{251} have also been shown to play critical roles in the oxidation (see below).

11.4 Reaction sequence

The overall mechanism is complex because of the multiple changes in the iron oxidation state. The hydroxylation step is thought to involve a "rebound" mechanism as shown below in Fig. 11.3.

$$FeO\cdot + R'_3CH \longrightarrow FeOH + R'_3C\cdot \longrightarrow Fe + R'_3COH$$

FIGURE 11.3 The "rebound" mechanism for hydroxylation of an unactivated C-H bond.

11.5 Detailed mechanism and the role of the active site residues

A detailed mechanism is shown in Fig. 11.4. Though several of the intermediates have been observed, the exact movement of electrons as the mechanism proceeds is not definitive. The scheme is constructed, and intermediates are presented, so as to preserve charge throughout the sequence. The heme is depicted schematically, with the "P" representing the porphyrin pi system.

The reaction begins with a "low spin" Fe(III) atom bound weakly to a water molecule. This water is displaced upon entry of the camphor into the active site (step a); the orientation of the camphor is directed by hydrogen bonding to Tyr_{96}. The entry of the substrate also alters the conformation of the enzyme from an "open" form to a "closed" form. The now pentacoordinate ferric atom converts to a "high spin" electronic configuration which makes it more readily reduced (i.e., makes the reduction potential more positive). The reduction of the ferric atom to the ferrous oxidation level occurs upon complexation with Pdx on the Cys ligand side of the heme (step b). Arg_{112} is important for the affinity of reduced Pdx for the ferric cytochrome (a key interaction between the two proteins is between Arg_{112} and the Pdx Asp38). It is also proposed to be part of the electron-transfer route between the two proteins; its replacement by other residues affects the oxidation–reduction potential of the heme iron. (Note that the oxidized Pdx is subsequently reduced by Pdx reductase (Pdr; EC 1.18.1.3) which contains the FAD and NAD cofactors). The electron transfer step has been postulated to occur by both a through-space interaction between the metal atoms and through-bond electron transfer. Only after the reduction, the iron can bind to oxygen (step c) to create an oxyferrous intermediate that is best represented as a ferric-bound superoxide. A further reduction (step d) forms a peroxo intermediate, which is then protonated (step e) to give a hydroperoxo species. A second protonation and concomitant loss of water (step f) generate the ferryl [Fe(IV)] species (see also Chapter 20: Indoleamine 2,3-Dioxygenase-1, and Chapter 25: Non-Heme Iron Halogenase) which can be represented as a resonance hybrid of an oxidized porphyrin ring system (i.e., $P\cdot^{+}$), and a second contributing structure with an uncharged P and an odd electron on the oxygen. The "rebound" mechanism shown in Fig. 11.2 is then invoked to rationalize product formation; hydrogen abstraction (step g) leads to a Fe(IV)/hydroxide complex that can also be represented as a Fe(III)/hydroxyl radical complex. Bonding of the camphor radical to this oxygen (step h) and then displacement by water (step i) leads to the product and restores the original iron complex.

The protonation steps ("e" and "f") in Fig. 11.4 have been examined in considerable detail. Asp_{251} is central to this chemistry. It facilitates the positioning of the Thr_{252} (see below), the protonation of the distal oxygen of the iron-bound oxygen by a water molecule (WAT901) that has moved into the active site, and the "shuttling" of protons from the bulk solvent via a series of additional, ordered water molecules. Asp_{251} can play this role only after conformational changes that occur after the complexation of CYP101A1 with Pdx. The essential role of Thr_{252} is attributed to H-bonding of the distal OH of the hydroperoxide to the oxygen atom of the Thr OH group. This interaction minimizes a competitive cleavage of the hydroperoxide to hydrogen peroxide.

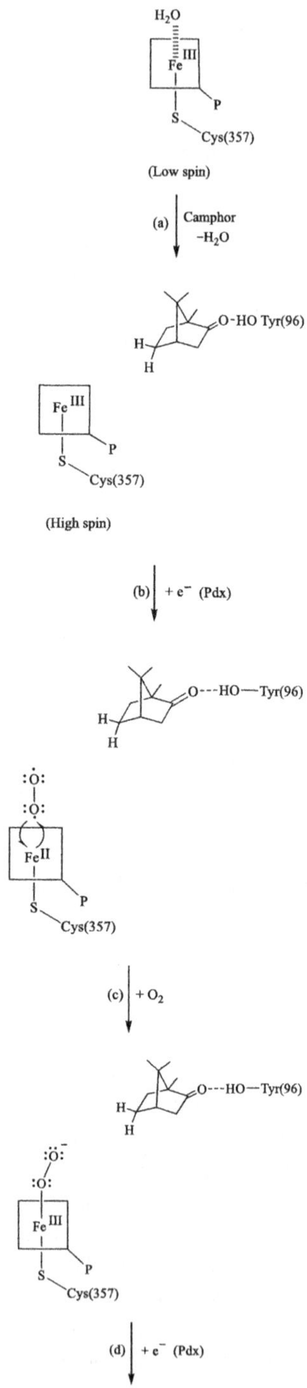

FIGURE 11.4 Detailed mechanism for CYP101A1.

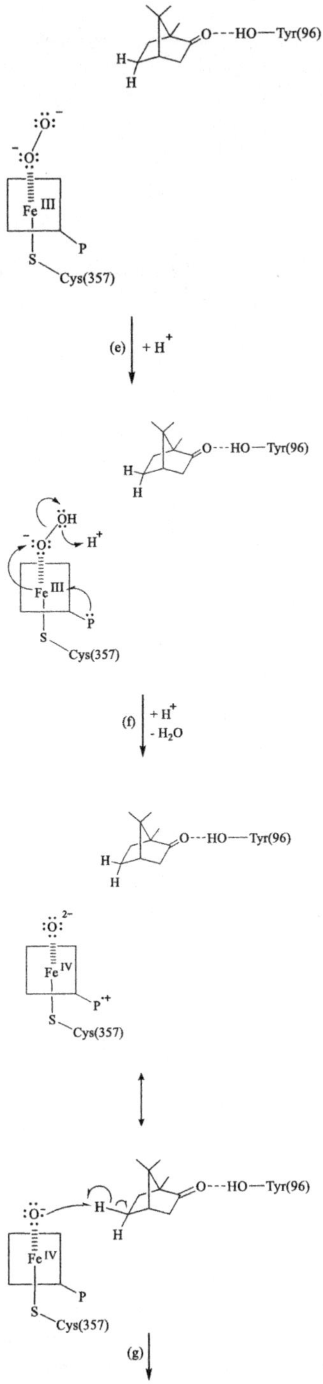

FIGURE 11.4 (Continued)

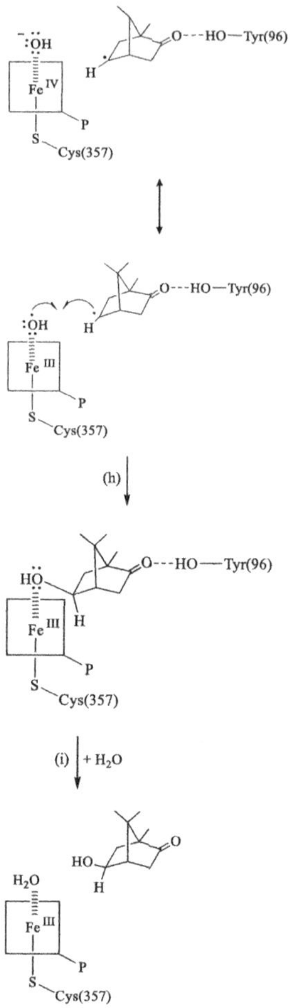

FIGURE 11.4 (Continued)

Leading references

Huang, X. and Groves, J. T. *Chem. Rev.* **118**, 249–253 (2017); Cytochrome P450. Structure, mechanism and biochemistry. P.R. Ortiz de Montellano, Ed. 4th Edition, Springer, 2015; Skinner, S. P.; Liu, W.-M.; Hiruma, Y.; Timmer, M.; Blok, A.; Hass, M. A. S.; Ubbink, M. *Proc. Natl. Acad. Sci. U. S. A.* **112**, 9022–9027 (2015); Poulos, T. L. *Chem. Rev.* **114**, 3919–3962 (2014); Rittle, J. and Green, M. T. *Science* **330**, 933–937 (2010); Nagano, S. and Poulos, T. L. *J. Biol. Chem.* **280** 31659-31653 (2005); Shimada, H.; Nagano, S.; Hori, H.; Ishimura, Y. *J. Inorg. Biochem.* **83**, 255–260 (2001); Unno, M.; Shimada, H.; Toba, Y.; Makino, R.; Ishimura, Y. *J. Biol. Chem.* **271**, 17869–17874 (1996); Kimata, Y.; Shimada, H.; Hirose, T.; Ishimura, Y. *Biochem. Biophys. Res. Commun.* **208**, 96–102 (1995); Poulos, T. L.; Finzel, B. C.; Howard, A. J. *J. Mol. Biol.* **195**, 687–700 (1987).

Chapter 12

m^5C Cytosine methyltransferase

12.1 m^5C Cytosine methyltransferase

m5C Cytosine methyltransferase (EC 2.1.1.37; C-5 cytosine-specific DNA methylase; m^5C-MTase) is one of several enzymes that methylate cytosine or adenine in DNA. It specifically methylates the C-5 carbon of cytosine; the others methylate the N^4 position of cytosine and the N^6 position of adenine. m^5C is the most common methylase in mammals and is labeled DNMT1 (DNA methyltransferase 1). However, much of the structural and mechanistic information about this class of enzymes is derived from studies on a bacterial analog, M.HhaI, derived from *Haemophilus haemolyticus*. There is considerable conservation of the AA's at the active site between these two (and with m^5C-MTases in general) and the mechanisms for the two methylases are quite similar. M.HhaI recognizes the 5′-GCGC-3′ sequence and methylates the underlined cytosine.

The requisite cofactor for this reaction is a ubiquitous methylating agent, *S*-adenosyl-L-methionine (AdoMet). (AdoMet is formed by the displacement of inorganic phosphate from ATP by the sulfur atom of methionine). The demethylation of AdoMet leads to *S*-adenosyl-L-homocysteine (AdoHcy; see Figs. 12.1 and 12.2).

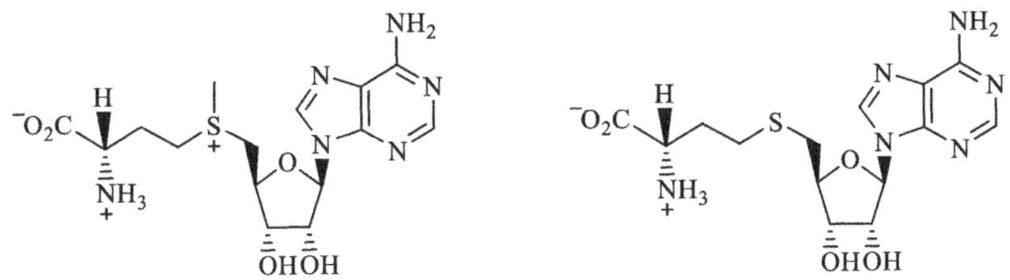

AdoMet AdoHcy

FIGURE 12.1 Structures of AdoMet and AdoHcy.

Enzyme Active Sites and their Reaction Mechanisms. DOI: https://doi.org/10.1016/B978-0-12-821067-3.00012-X

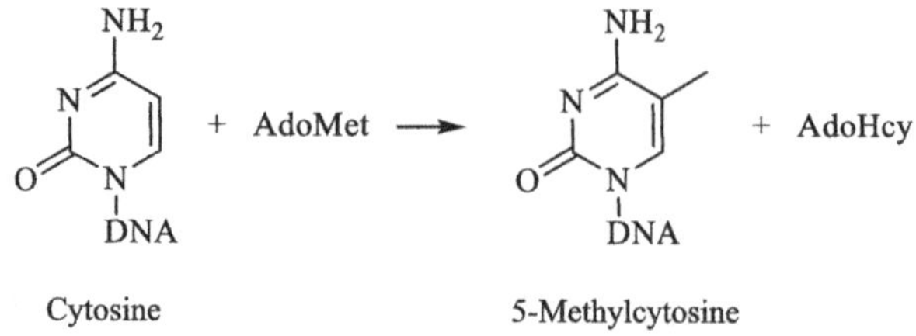

FIGURE 12.2 Overall reaction of m^{5}C-MTase.

12.2 Physiological function

DNA methylation is critical for embryogenesis and, in general, for normal cellular function. DNMT1, but not M.HhaI, shows a strong preference for DNA that is already partially methylated. Bacteria employ methylation to avoid self-degradation by their restriction endonuclease enzymes.

12.3 Key structural features

A critical feature of this chemistry is the "flipping" of the target cytosine out of the DNA helix. (Another example of such "flipping" may be found in the mechanism of repair by DNA photolyase). This serves to place the cytosine into the active site pocket of the enzyme. The active site for M.HhaI is shown in Fig. 12.3. The role of Arg_{163} and Arg_{165} has been ascribed to electrostatic activation of the cytosine ring toward nucleophilic attack and the

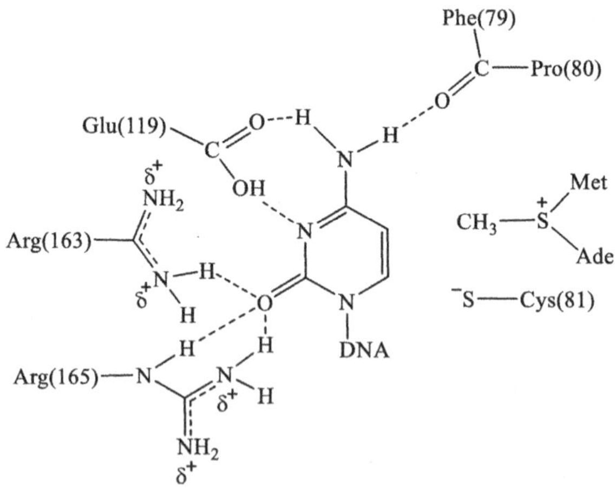

FIGURE 12.3 The active site of M.HhaI showing key amino acid residues. *Adapted with permission from Aranda et al.* ACS Catal. *6, 3262–3276 (2016); Copyright 2016 American Chemical Society.*

possible (see below) stabilization of the initial intermediate. Arg_{165} is also thought to facilitate the base "flipping" and to help position the target base for attack by the cysteine (see below). Glu_{119} has been invoked as a proton donor. It also serves to activate the cytosine and possible stabilize an intermediate. Finally, Cys_{81} initiates the chemistry by adding to the cytidine double bond at the C6 position. These primary, and heavily conserved, residues are shown in Fig. 12.3. Several water molecules found within the active site have been omitted for clarity.

12.4 Reaction sequence

The methyl transfer chemistry has classically been viewed as a consequence of three chemical steps: (1) the addition of a nucleophile ($:N^-$) to a conjugated double bond to form a resonance-stabilized anion, (2) nucleophilic attack by the anion on an electrophile (CH_3) with the expulsion of a good leaving group (LG^+), and (3) base-catalyzed deprotonation with the elimination of the initial nucleophile to regenerate the double bond. The nucleophile thus plays the role of a catalyst in this chemistry. These steps are illustrated in Fig. 12.4. As will be seen below, the chemistry is not so straightforward.

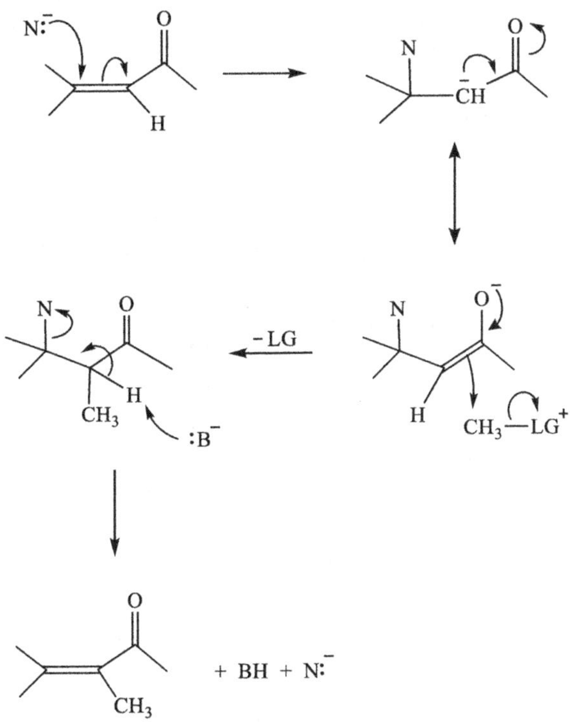

FIGURE 12.4 A classic view of the reaction sequence for methyltransferase chemistry.

12.5 Detailed mechanism(s) and the role of the active site residues

A proposed mechanism is shown in Fig. 12.5. The chemistry is initiated in step (a) by the nucleophilic, conjugate addition of the thiolate anion (the

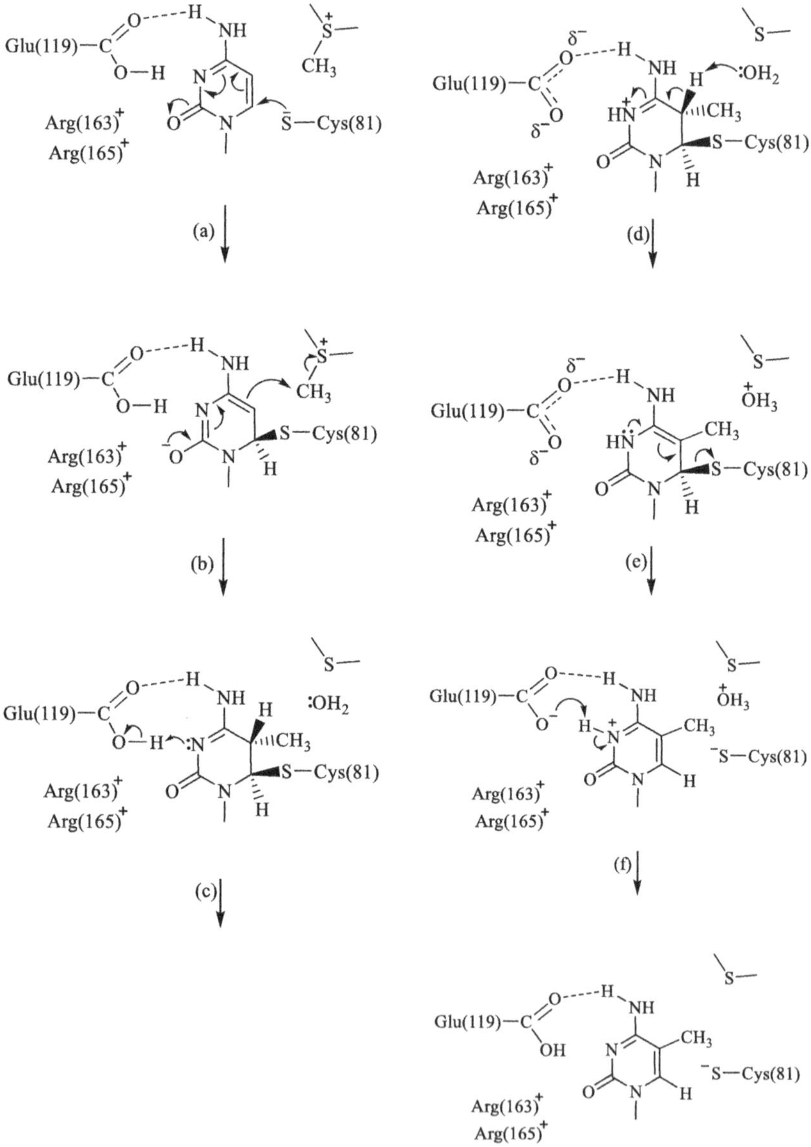

FIGURE 12.5 Proposed mechanism for the methyltransferase reaction. See Fig. 12.3. *Copyright 2016 American Chemical Society.*

state of protonation of this sulfur atom has been a subject of debate) on the cytosine ring to form an extensively resonance-stabilized oxyanion. There is a proposal that the sulfur is deprotonated by a DNA phosphate group (see Zangi et al., 2010). In step (b), the anion attacks the electrophilic, methylated sulfonium ion—importantly, from the alkene face opposite to the initial addition (the methyl transfer step is thought to be rate limiting). Because of this, the regeneration of the cytosine double bond, if concerted, would be a disfavored, high-energy *syn* elimination reaction. A step-wise process is therefore invoked in the following steps. The ring is protonated by Glu_{119} in step (c) and an active site water is invoked for the deprotonation in step (d). Heterolytic cleavage of the C-S bond (step e) and proton transfer back to the Glu_{119} (step f) complete the reaction.

Leading references

Aranda, J.; Zinovjev, K.; Śwideerek, K.; Roca, M.; Tuñón, I. *ACS Catal.* **6**, 3262–3276 (2016); Zhang, Z.-M.; Liu, S.; Lin, K.; Luo, Y.; Perry, J. J.; Wang, Y.; Song, J. *J. Mol. Biol.* **427**, 2520–2531 (2015); Zangi, R.; Arrieta, A.; Cossio, F. P. *J. Mol. Biol.* **400**, 632–644 (2010); Youngblood, B.; Shieh, F.-K.; Buller, F.; Bullock, Y.; Reich, N. O. *Biochemistry* **46**, 8766–8775 (2007); Shieh, F.-K. and Reich, N. O. *J. Mol. Biol.* **373**, 1157–1168 (2007); Zhang, X. and Bruice, T. C. *Proc. Natl. Acad. Sci. U. S. A.* **103**, 6148–6153 (2006).

Chapter 13

Deoxyribodipyrimidine photolyase

13.1 Deoxyribodipyrimidine photolyase

Deoxyribodipyrimidine photolyase (EC 4.1.99.3; CPD DNA photolyase; photoreactivating enzyme; Phr1) catalyzes the repair of light-damaged DNA using near-UV and visible (to 500 nm) light (termed "photoreactivation"). The reaction is very efficient; it proceeds with a quantum efficiency of 0.82. The vast majority (c. 80%) of lesions in DNA caused by exposure to UV (particularly 290–320 nm; "UVB") light are intra-strand cyclobutane pyrimidine dimers (CPD). Pyrimidine-pyrimidone (6-4) lesions constitute the remainder. Thus photolyases can be subdivided based on whether they repair the CPD or the (6-4) lesions. CPD photolyases have been further grouped into three classes: I, II, and III. Those found in bacteria are members of class I and it is for these that the greatest mechanistic detail is known. This enzyme requires two cofactors, flavin adenine dinucleotide (FAD) and either 5,10-methenyltetrahydrofolylpolyglutamate (MTHF) or 7,8-didemethyl-8-hydroxy-5-deazariboflavin (8-HDF). The FAD is active in its reduced form, $FADH^-$, and electron transfer from it to the CPD initiates the sequence for repair. MTHF and 8-HDF are light-harvesting antenna molecules that help in the absorption of the requisite photons for this light-driven chemistry. The four possible pyrimidine dimers are 5′-TT-3′, 5′-TC-3′, 5′-CT-3′, and 5′-CC-3′. Of these, the most readily formed is T-T. The overall process is called nucleotide excision repair (NER). See also "base excision repair" in Chapter 39, Uracil-DNA Glycosylase. The conversion of thymine dinucleotide into the corresponding CPD is shown in Fig. 13.1. The structures of FAD, $FADH^-$, MTHF, and 8-HDF are given in Figs. 13.2–13.5.

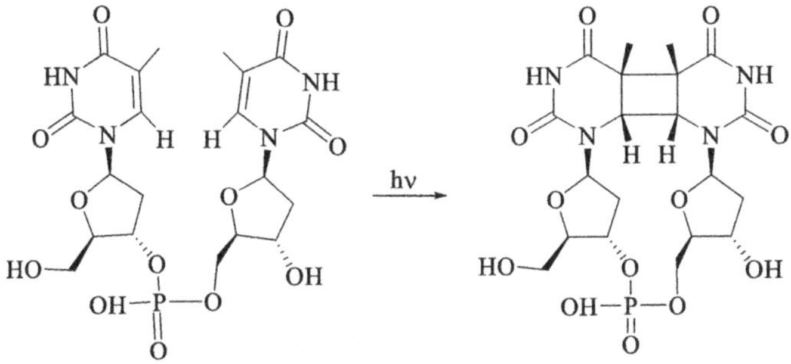

FIGURE 13.1 Photolytic conversion of TpT into a CPD. *TpT*, Thymine dinucleotide; *CPD*, cyclobutane pyrimidine dimers.

Enzyme Active Sites and their Reaction Mechanisms. DOI: https://doi.org/10.1016/B978-0-12-821067-3.00013-1

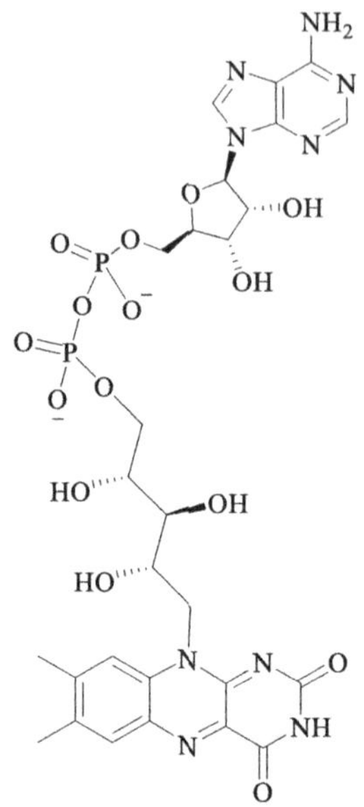

FIGURE 13.2 Structure of FAD. *FAD*, Flavin adenine dinucleotide.

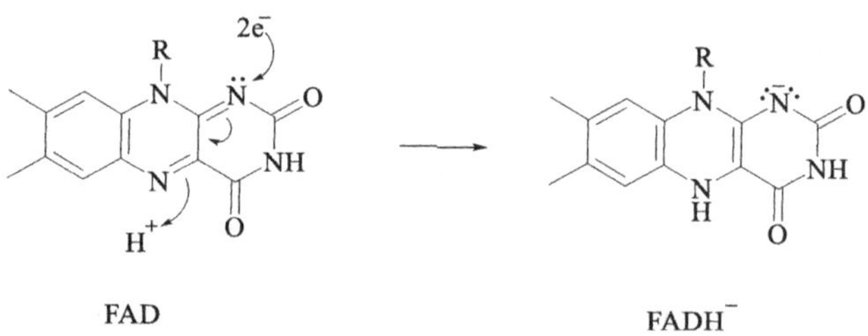

FIGURE 13.3 Structure of FADH$^-$ showing its relationship to FAD. *FAD*, Flavin adenine dinucleotide.

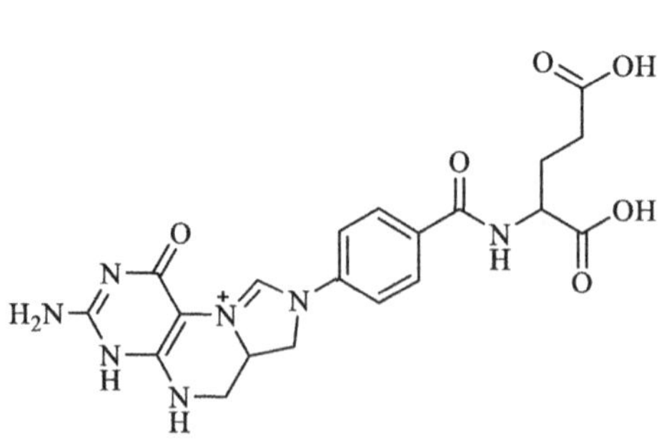

FIGURE 13.4 Structure of MTHF. *MTHF*, 5,10-Methenyltetrahydrofolylpolyglutamate.

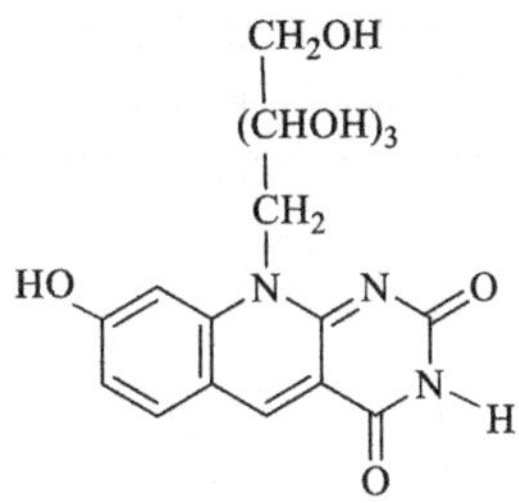

FIGURE 13.5 Structure of 8-HDF. *8-HDF*, 7,8-Didemethyl-8-hydroxy-5-deazariboflavin.

13.2 Physiological function

Damage to DNA by sunlight leads to mutagenesis and eventually to cancer. Thus enzymes that repair damage to DNA caused by sunlight are ubiquitous. Placental animals (including humans) lack photolyase but instead rely on NER for this task.

13.3 Key structural features

Photolyases are monomeric proteins consisting of approximately 500 AA residues. FAD lies within the active site, 3.1 Å from the lesion, but separated from the second cofactor by 9 Å (*Aspergillus nidulans* with 8-HDF) to 17 Å (*Escherichia coli* with MTHF). Thus the transfer of energy from the cofactor to FAD is attributed to long-range resonance energy transfer. An unusual feature is that the FAD is bound in a U-shape so that the adenine ring lies close to the flavin group. This orientation allows the adenine to play a role in the electron transfer between the FADH (see below). Photolyase does not bind to any specific DNA sequence but rather recognizes the CPD, which is "flipped" out of the double helix and into the active site.

The X-ray structure for the active site of the *A. nidulans* photolyase bound to a 14-base nucleotide containing a pair of thymine residues that form a CPD is shown in Fig. 13.6. Asn_{386} hydrogen bonds to the N5 hydrogen of reduced FAD and is also presumed to help stabilize the FAD radical (not shown). In *E. coli*, the reduction of excited FAD (the first step in the photolyase mechanism) involves an electron transfer chain that includes a highly conserved tryptophan "triad" (Trp_{382}, Trp_{359}, and Trp_{306}). Trp_{306} is furthest away from the FAD and lies on the protein surface where it can be reduced by an exogenous reductant. The favorable energetics of this electron transfer sequence is due to the dependence of the oxidation potentials of the

Trp residues on the degree to which the Trp radical cations are exposed to, and therefore stabilized by, a polar, hydrophilic environment. Trp_{316} and Trp_{384} have also been shown to contribute to the reduction of the FAD.

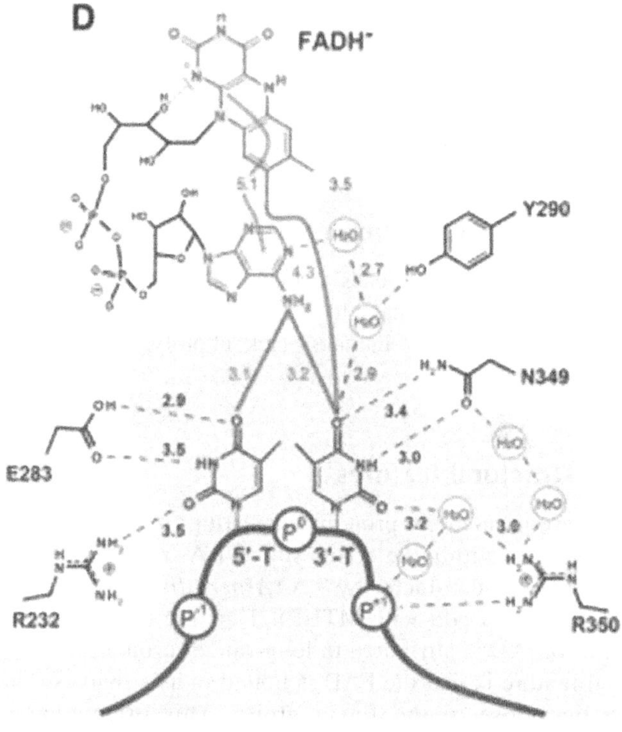

FIGURE 13.6 The active site of *A. nidulans* containing the in situ repair product of a CPD lesion. The location of the adenine portion of the FADH$^-$ is noteworthy. *CPD*, Cyclobutane pyrimidine dimer. *Republished with permission from the American Association for the Advancement of Science, from Mees et al.* Science ***306****, 1789–1703 (2004); Copyright 2004 Copyright Clearance Center, Inc.*

13.4 Reaction sequence

The general chemical sequence for photolyase is shown schematically in Fig. 13.7. The cleavage of the CPD is initiated [step (a)] by an electron transfer from an anionic (D$^-$) reductant to form the radical, D$^\bullet$, and a CPD radical anion. The cyclobutane bonds of this radical anion then cleave (steps b and c) to regenerate the two original pyrimidine double bonds. The final step (step d) is the back-transfer of an electron to D$^\bullet$ to regenerate D$^-$.

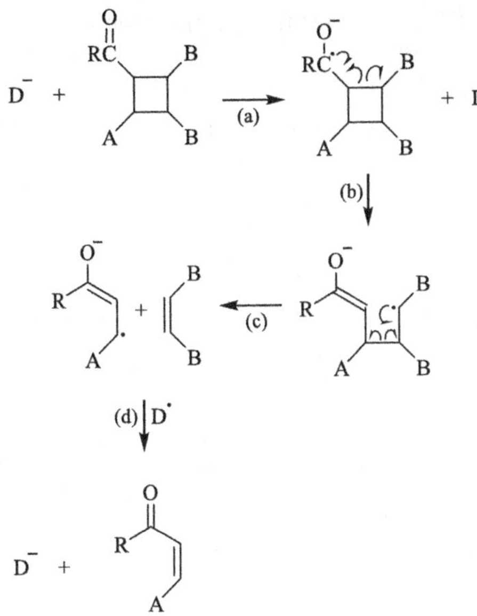

FIGURE 13.7 Cleavage of a cyclobutane into two alkenes by initial reduction to a radical anion.

13.5 Detailed mechanism and the role of active site residues

The primary roles of the active site residues are to help hold the reaction participants in place and stabilize reaction intermediates. Of particular note is a proposal for a "transient proton transfer" by Glu_{274}/Glu_{283} to the CPD radical anion. The complete mechanism for photolyase is given in Fig. 13.8. In step (a) FAD is photoreduced by the Trp residues of the enzyme to form the requisite $FADH^-$. The $FADH^-$ electronic excited state is then formed, either by direct absorption of light [step (b)] or by energy transfer from the excited state of one of the antenna cofactors [step (c)]. Electron transfer from the $FADH^-$ to the CPD follows in step (d) to form $FADH^\bullet$ and the CPD radical anion. The electron is thought to originate from the isoalloxazine chromophore; the adenine ring, which lies between the isoalloxazine and the CPD due to the bent FAD conformation, facilitates its transfer. There is also good evidence that the electron preferentially goes to the 5′ side of the dimer. Homolytic cleavage of the C5/C5 bond occurs in step (e) (the intermediate is oft depicted as shown on the right side), to be followed by heterolytic cleavage of the C6/C6 bond in step (f). This forms the 5′-thymine product and a 3′-thymine radical anion. Back electron transfer to the $FADH^\bullet$ completes the repair and regenerates $FADH^-$. As with the initial electronic transfer, the adenine moiety of the FAD seems to facilitate this process.

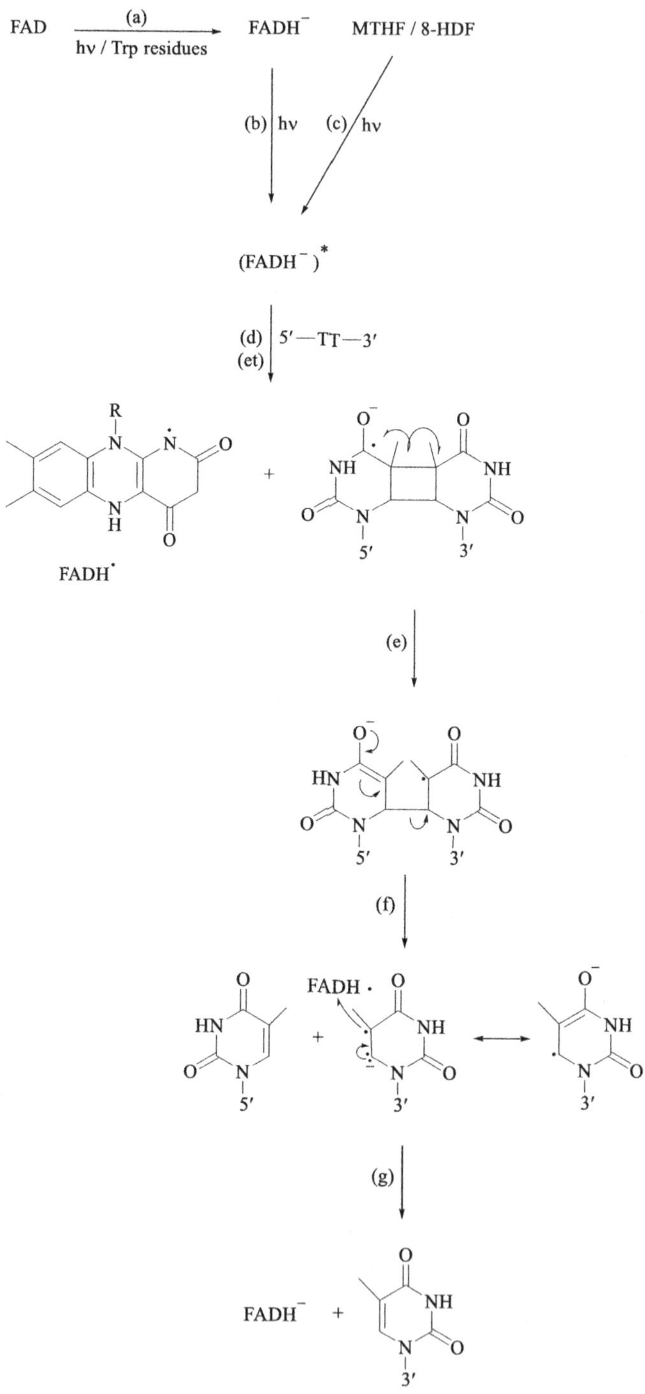

FIGURE 13.8 Mechanism for photolyase.

Leading references

Sato, R.; Kito-Nishioka, H.; Ando, K.; Yamato, T. *J. Phys. Chem.* **B8**, 6912–6921 (2018); Liu, Z.; Wang, L.; Zhong, D. *Phys. Chem. Chem. Phys.* **17**, 11933–11949 (2015); Tan, C.; Liu, Z.; Li, J.; Guo, X.; Wang, L.; Sancar, A.; Zhong, D. *Nat. Commun.* **6**, 7302–7307 (2015); Kneuttlinger, A. C.; Kashiwazake, G.; Prill, S.; Heil, K.; Müller, M.; Carell, T. *Photochem. Photobiol.* **90**, 1–14 (2014); Liu, Z.; Tan, C.; Guo, X.; Li, J.; Wang, L.; Sancar, A.; Zhong, D. *Proc. Natl. Acad. Sci. U. S. A.* **110**, 12966–12971 (2013); Liu, Z.; Tan, C.; Guo, X.; Kao, Y.-T.; Li, J.; Wang, L.; Sancar, A.; Zhong, D. *Proc. Natl. Acad. Sci. U. S. A.* **108**, 14831–14836 (2011); Müller, P.; Ignatz, E.; Klontle, S.; Brettel, K.; Essen, L.-O. *Chem. Sci.* **9**, 1200–1212 (2008); Mees, A.; Klar, T.; Gnau, P.; Hennecke, U.; Eker, A. P. M.; Carell, T.; Essen, L.-O. *Science* **306**, 1789–1793 (2004).

Chapter 14

Dihydrolipoamide dehydrogenase

14.1 Dihydrolipoamide dehydrogenase

Dihydrolipoamide dehydrogenase (EC 1.8.1.4; E3; dihydrolipoyl dehydrogenase; dihydrolipoamide: NAD^+ oxidoreductase) is one enzymatic component of the mitochondrial-based pyruvate dehydrogenase multienzyme complex (PDHC; PDC; PDH). This complex is an associated set of three enzymes that ultimately converts pyruvate to acetyl coenzyme A (acetyl-CoA) (see below). Other components of the PDHC are pyruvate dehydrogenase (E1p, EC 1.2.4.1) and dihydrolipoyl transacetylase (E2p, EC 2.3.1.12). E3 is the third component of the PDHC and is found in all α-ketoacid dehydrogenase complexes. Its role is to oxidize the dihydrolipoamide-E2p back to the corresponding disulfide. As seen below, E3 is ultimately reoxidized to the disulfide. Flavin adenine dinucleotide (FAD; see Chapter 13: Deoxyribodipyrimidine photolyase for the structure of FAD) and nicotinamide adenine dinucleotide (NAD^+; see Fig. 14.7 below) are requisite cofactors. The overall chemistry catalyzed by E3 is shown in Fig. 14.1.

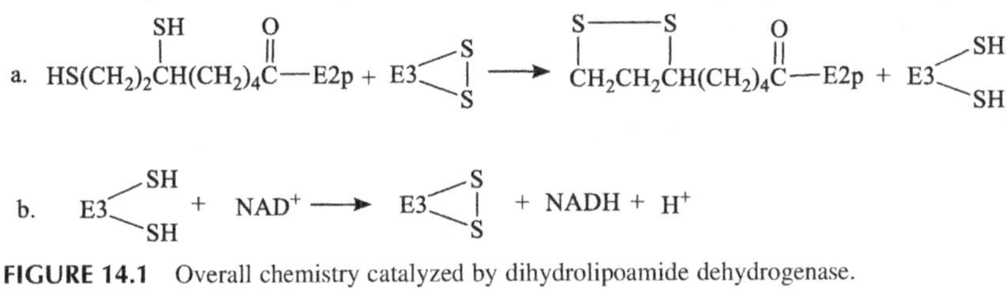

FIGURE 14.1 Overall chemistry catalyzed by dihydrolipoamide dehydrogenase.

14.2 Physiological function

Acetyl-CoA is utilized in the citric acid cycle, so that the PDHC serves to link glycolysis to this key component of cellular respiration. The net consequence of this series of reactions is shown below in Fig. 14.2.

Enzyme Active Sites and their Reaction Mechanisms. DOI: https://doi.org/10.1016/B978-0-12-821067-3.00014-3

$$\underset{\text{(Pyruvate)}}{CH_3\overset{O}{\overset{\|}{C}}CO_2^-} + CoASH + NAD^+ \longrightarrow \underset{\text{(Acetyl-CoA)}}{CH_3\overset{O}{\overset{\|}{C}}SCoA} + CO_2 + NADH$$

FIGURE 14.2 Conversion of pyruvate to acetyl-CoA. *Acetyl-CoA*, Acetyl coenzyme A.

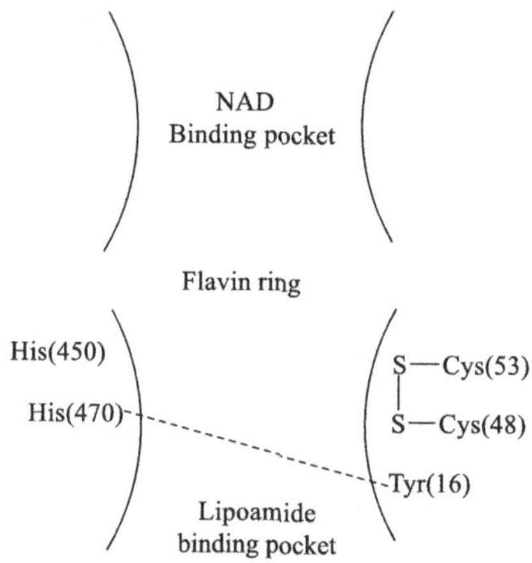

FIGURE 14.3 Schematic drawing of the active site of *Pseudomonas fluorescens. Adapted with permission from Mattevi et al.* J. Mol. Biol. ***230****, 1200–1215 (1993).*

In more detail, E1p converts pyruvate to enzyme-bound hydroxyethylidene-TPP and catalyzes its transfer of an acetyl moiety to "lipoamide-E2p" (lipoic acid covalently bound via an amide linkage to a lysine unit of E2p). This results in the formation of S-acetyldihydrolipoamide-E2p and reactivated E1p. E2p then transfers the acetyl group to coenzyme A with the concomitant formation of dihydrolipoamide-E2p.

14.3 Key structural features

The general arrangement of the cofactors and the lipoamide substrate within the active site is shown in Fig. 14.3. NAD^+ and FAD bind into separate domains such that the two aromatic rings are coplanar so as to maximize π-π interaction, with the NAD on the *re* side of the isoalloxazine ring. Tyr_{181} plays an interesting role in first shielding the FAD cofactor (primarily from premature oxidation of reduced FAD formed during the catalytic cycle) and then swinging out of the way, so that the reduced FAD can undergo oxidation/reduction chemistry with NAD^+. His_{450} is involved in the reaction mechanism (see below); a hydrogen bond bridging His_{470} and Tyr_{16} is noteworthy. More detailed representations of FAD and NAD^+ bound into the E3 active site of *A. vinelandii* and *P. putida*, respectively, are presented in Figs. 14.4 and 14.5.

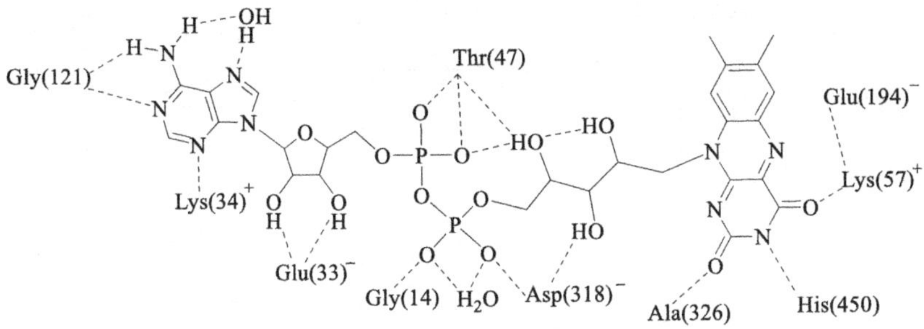

FIGURE 14.4 The FAD binding site in *Azotobacter vinelandii*. *FAD*, Flavin adenine dinucleotide. *Adapted with permission from Mattevi et al.* J. Mol. Biol. ***220****, 975–994 (1991).*

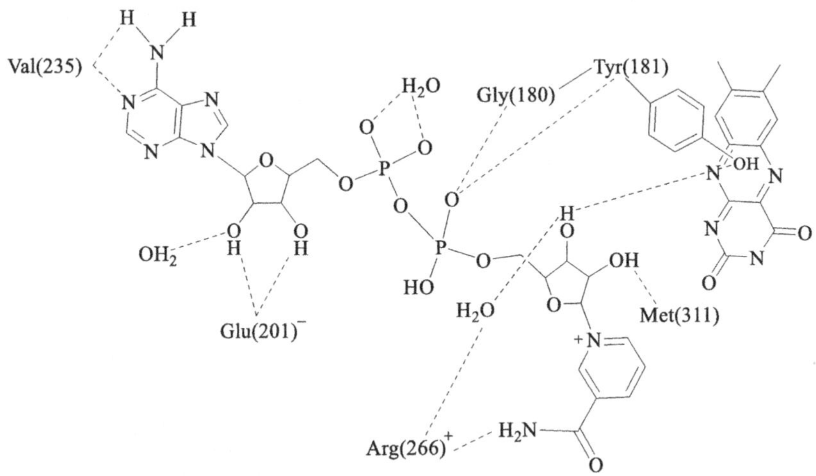

FIGURE 14.5 The NAD^+ binding site in *Pseudomonas putida*. *NAD^+*, Nicotinamide adenine dinucleotide. *Adapted with permission from Mattevi et al.* Proteins ***14****, 336–351 (1992); Copyright 1992 John Wiley and Sons.*

14.4 Reaction sequence

The reoxidization of the dihydrolipoamide is accomplished by a concomitant reduction of a disulfide linkage within E3. A summary of the E2p/E3 reaction sequence is shown in Fig. 14.6. The conversion of NAD^+ to NADH is shown in Fig. 14.7.

14.5 Detailed mechanism and the role of the active site residues

A detailed mechanism for the E3-catalyzed chemistry has been proposed using the X-ray structure for *P. putida*. Here Glu_{456} and His_{451} facilitate the deprotonation of, and subsequent nucleophilic attack by, dihydrolipoamide-E2p

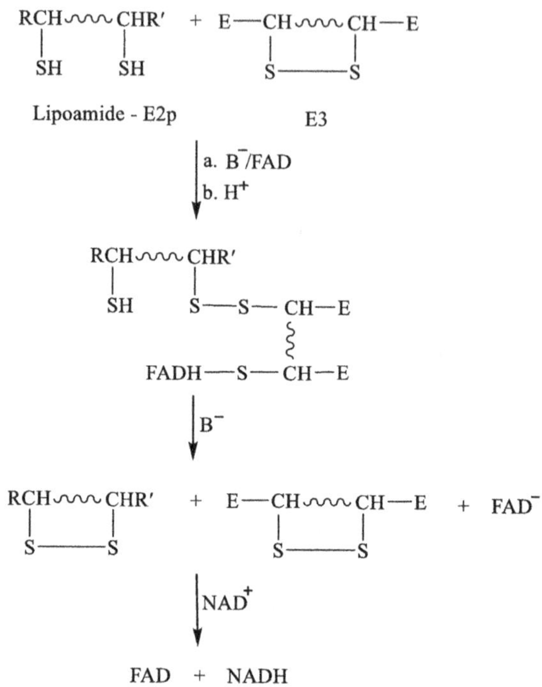

FIGURE 14.6 The reaction sequence for the oxidation of reduced E2p by E3.

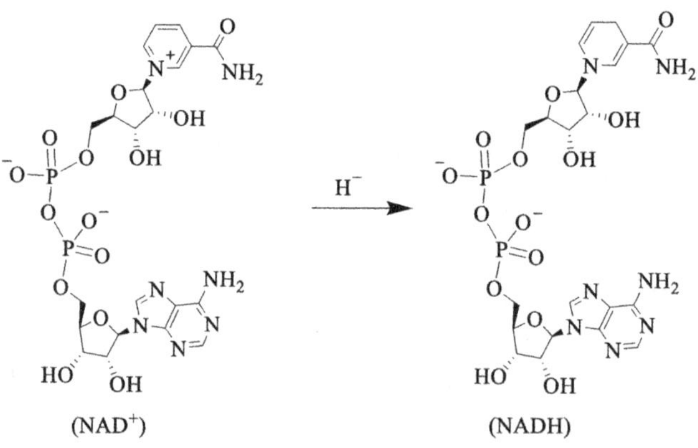

FIGURE 14.7 General reaction for the conversion of NAD^+ to NADH by the transfer of a hydride anion.

on an enzymatic disulfide linkage formed between Cys_{43} and Cys_{48} (Cys_{45} and Cys_{50} in the human enzyme). Tyr_{181} plays an interesting role in first shielding the FAD cofactor (primarily from premature oxidation of reduced FAD formed during the catalytic cycle) and then swinging out of the way, so that the reduced FAD can undergo oxidation/reduction chemistry with NAD^+. The terminal step involves an oxidation/reduction reaction involving reduced FAD and NAD^+.

In Fig. 14.8, step (a) involves the binding of dihydrolipoamide-E2p to the E3 active site. The FAD cofactor is also present and shielded by Tyr_{181}. In step (b) His_{451} deprotonates a thiol proton from the reduced disulfide, dihydrolipoamide-E2p, to initiate an attack on the E3 disulfide linkage

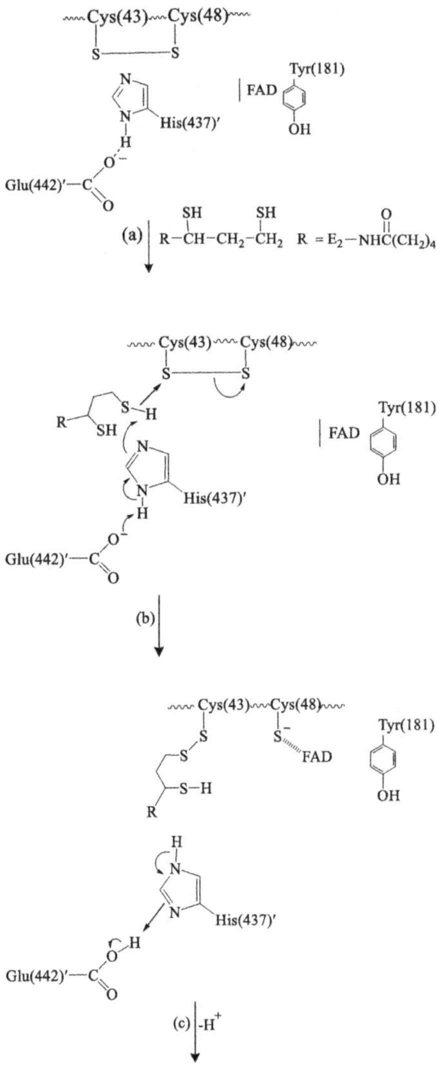

FIGURE 14.8 Detailed mechanism for E3p oxidation of dihydrolipoamide showing the involvement of key amino acids at the active site.

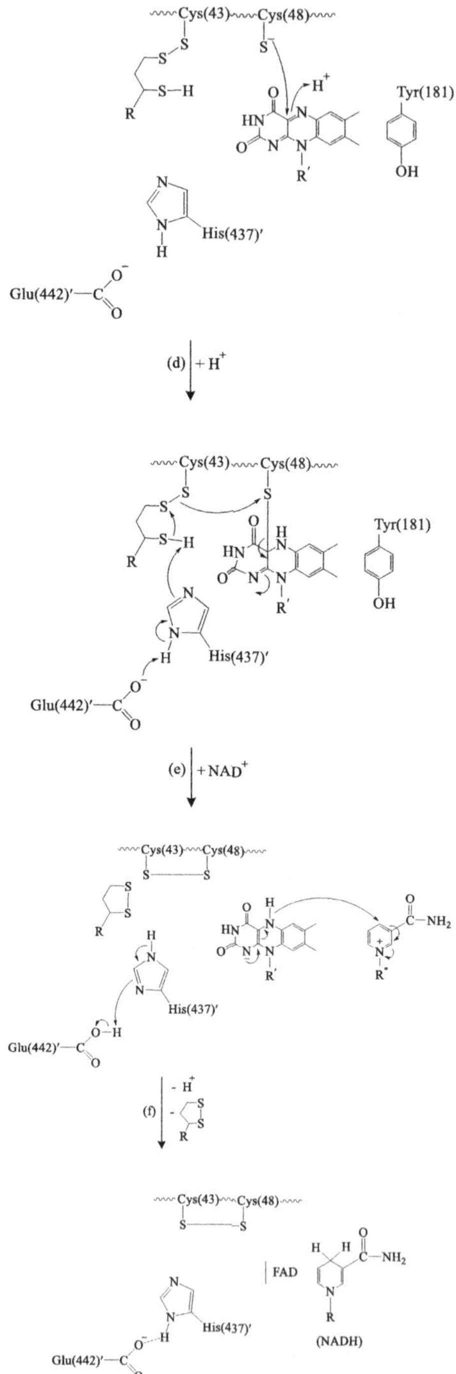

FIGURE 14.8 (Continued).

involving Cys_{43} and Cys_{48}. The deprotonation is facilitated by concomitant deprotonation of His_{451} by Glu_{456}. The result is the creation of a new disulfide linkage and a charge-transfer complex of the Cys_{48} anion with FAD. The Glu_{456} anion is regenerated in step (c) and the thiolate anion attacks FAD to form a thioether (step d). In step (e), His_{451} deprotonates the remaining thiol group on the E3-bound lipoamide (again facilitated by Glu_{456}), which leads to cleavage of the newly formed disulfide bond to generate the reoxidized lipoamide-E2p. A concomitant attack by the released Cys_{43} sulfur atom on the Cys_{48} sulfur linkage regenerates the E3 disulfide and forms $FADH^-$. At this point NAD^+ is inserted into the active site, with Tyr_{181} pushed aside, and a hydride anion is transferred from $FADH^-$ (step f) to form NADH and the FAD cofactor.

Leading references

Szabo, E.; Mizsei, R.; Wilk, P.; Zambo, Z.; Torocsik, B.; Weiss, M. S.; Adam-Vizi, V.; Ambrus, A. *Free Radic. Biol. Med.* **124**, 214–220 (2018); Wang, Y.-C.; Wang, S.-T.; Li, C.; Chen, L.-Y.; Liu, W.-H.; Chen, P.-R.; Chou, M.-C.; Liu, T.-C. *J. Biochem. Med. Sci.* **15**, 37–46 (2008); Ciszak, E. M.; Makal, A.; Hong, Y. S.; Vettalkkorumakankauv, A. K.; Korotchkina, L. G.; Patel, M. S. *J. Biol. Chem.* **38**, 648–655 (2006); Brautigam, C. A.; Wynn, R. M.; Chuang, J. L.; Machius, M.; Tomchick, D. R.; Chuang, D. T. *Structure* **14**, 611–623 (2006); Brautigam, C. A.; Chuang, J. L.; Tomchick, D. R.; Machius, M.; Chuang, D. T. *J. Mol. Biol.* **350**, 543–552 (2005); Kim, H. *J. Biochem. Mol. Biol.* **38**, 248–252 (2005); Argyrou, A.; Blanchard, J. S.; Palfey, B. A. *Biochemistry* **41**, 14580–14590 (2002); Mattevi, A.; Obmolova, G.; Kalk, K. H.; van Berkel, W. J. H.; Hol, W. G. J. *J. Mol. Biol.* **230**, 1200–1215 (1993); Mattevi, A.; Obmolova, G.; Sokatch, J. R.; Betzel, C.; Hol, W. G. J. *Proteins* **13**, 336–351 (1992); Jentoft, J. E.; Shoham, M.; Hurst, D.; Patel, M. S. *Proteins* **14**, 88–101 (1992); Mattevi, A.; Schierbeck, A. J.; Hol, W. G. J. *J. Mol. Biol.* **220**, 975–994 (1991).

Chapter 15

Dihydrolipoyl transacetylase

15.1 Dihydrolipoyl transacetylase

Dihydrolipoyl transacetylase (EC 2.3.1.12; E2p; dihydrolipoamide acetyltransferase) is one enzymatic component of the mitochondrial-based pyruvate dehydrogenase multienzyme complex (PDHC; PDC; PDH), an associated set of three enzymes that ultimately convert pyruvate to acetyl coenzyme A (acetyl-CoA). Other components of the PDHC are pyruvate dehydrogenase (E1p, EC 1.2.4.1) and dihydrolipoyl dehydrogenase (E3, EC 1.8.1.4). The role of E2p is to catalyze the transfer of an acetyl group from S-acetyldihydro-lipoamide-E2p to CoA. The structure of CoA is shown in Fig. 15.1 (note that CoA is a phosphate ester formed from pantetheine and 3′ phosphorylated adenosine diphosphate). The chemical conversion is shown in Fig. 15.2.

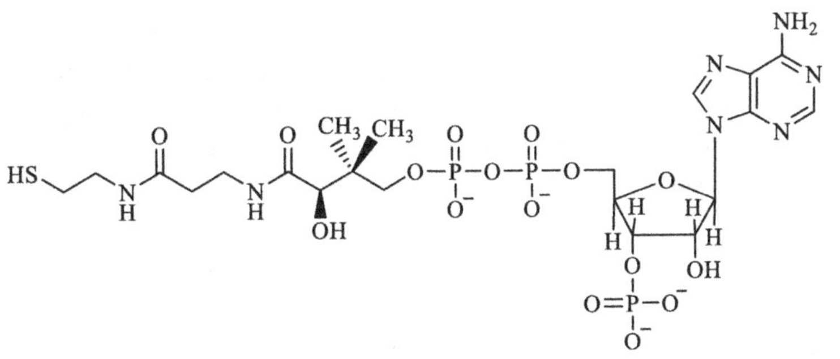

FIGURE 15.1 Coenzyme A.

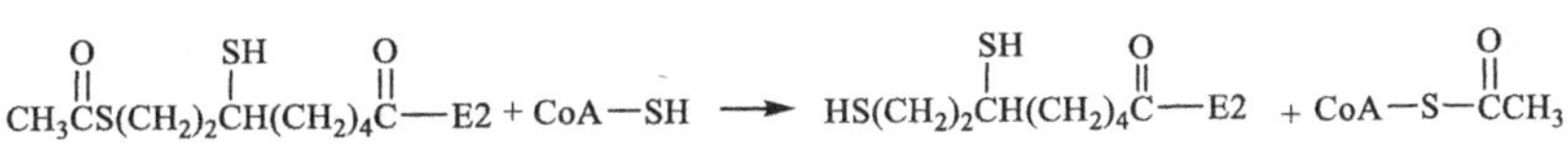

(S-acetyldihydrolipoamide-E2)

FIGURE 15.2 Overall chemistry catalyzed by dihydrolipoyl transacetylase.

Enzyme Active Sites and their Reaction Mechanisms. DOI: https://doi.org/10.1016/B978-0-12-821067-3.00015-5

15.2 Physiological function

Acetyl-CoA is utilized in the citric acid cycle, so that the PDHC serves to link glycolysis to this key component of cellular respiration. The net chemistry catalyzed by the PDHC is shown in Fig. 15.3.

$$\underset{\text{(Pyruvate)}}{CH_3\overset{\overset{\displaystyle O}{\|}}{C}CO_2^-} + CoASH + NAD^+ \rightarrow \underset{\text{(Acetyl-CoA)}}{CH_3\overset{\overset{\displaystyle O}{\|}}{C}SCoA} + CO_2 + NADH$$

FIGURE 15.3 The overall chemistry catalyzed by the PDHC.

In more detail, E1p converts pyruvate to enzyme-bound hydroxyethylidene-TPP and catalyzes its transfer of an acetyl moiety to "lipoamide-E2p" (lipoic acid covalently bound via an amide linkage to a lysine unit of E2p). This results in the formation of S-acetyldihydrolipoamide-E2p and reactivated E1p. E2p then transfers the acetyl group to CoA. The release of acetyl-CoA is accompanied by the formation of enzyme-bound dihydrolipoamide, which ultimately is oxidized by E3 to lipoamide. This is then available to E2p to continue the catalytic cycle.

15.3 Key structural features

Most of the structural information for E2p derives from the crystallographic analysis of the catalytic domain of bacterial E2p. Though no such is yet available for human E2p, information on the human E2p catalytic center has been obtained by cryo-electron microscopy. The catalytic center lies at the interface between two subunits with separate binding pockets for CoA and S-acetyldihydrolipoamide at the opposite ends of a channel. A His and a Ser are well-conserved essential residues at the catalytic site. Likewise, the residues that flank this His (Asp and Arg) are observed in all E2 active sites. These form a salt bridge which, it is proposed, helps to constrain the His into the appropriate tautomer and conformation that allows for intramolecular hydrogen bonding between the His amide carbonyl group and its ring NH (HD1) hydrogen atom (see Fig. 15.4 for *Azotobacter vinelandii*). This places the NE2 atom close to the SH group of CoA (see below), thus facilitating the presumed general-base function of this His residue (though that long-ascribed role has recently been questioned). The role of the conserved Ser is thought to be the stabilization of the oxyanion intermediate formed during the acetyl transfer chemistry. The residues important to the binding sites for CoA and S-acetyldihydrolipoamide have also been elaborated for *A. vinelandii*.

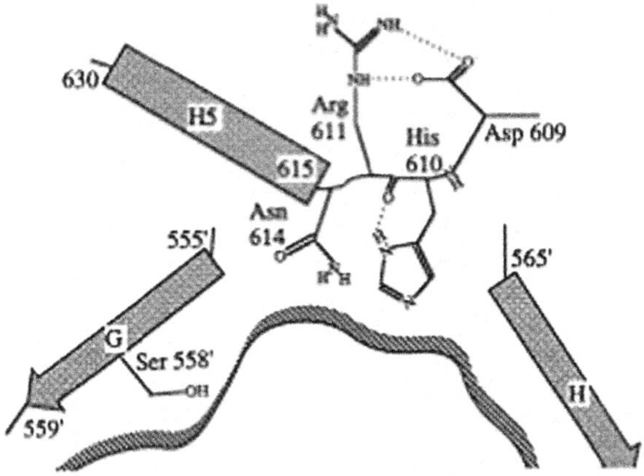

FIGURE 15.4 The Asp, Arg, His triad in the *A. vinelandii* active site. *Republished with permission of the American Association for the Advancement of Science from Mattevi et al., Science 255, 1544–1550 (1992); Copyright 1992 Copyright Clearance Center, Inc.*

15.4 Reaction sequence

This chemistry may be thought of as the thio equivalent of a transesterification reaction. The generalized base-catalyzed reaction is illustrated in Fig. 15.5 presuming the role of the histidine at the active site. Step (a) involves activation of the thiol by deprotonation to form the thiolate anion, step (b) is the nucleophilic attack of the thiolate on a thioester to form a tetrahedral intermediate (as in the analogous alcohol chemistry), and step (c) involves formation of the new thioester and the leaving group thiolate anion. See elsewhere in this book for the mechanisms of the chemistry catalyzed by E1p and E3.

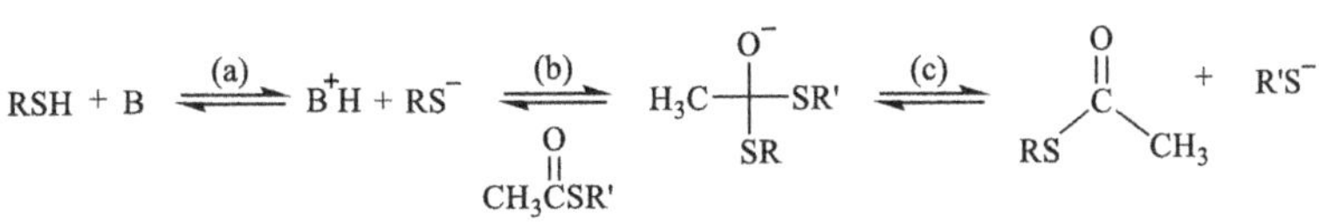

FIGURE 15.5 General reaction sequence for the thio equivalent of a transesterification reaction.

15.5 Detailed mechanism and the role of the active site residues

This is shown in Fig. 15.6 for *A. vinelandii*. Only the conserved His and Ser residues at the active site have been invoked as facilitating the transacetylation reaction. However, note also the proposed role of Asp and Arg in assuring the requisite His tautomer is in its optimal conformation (Fig. 15.4). In step (a) His_{610} deprotonates CoASH, which in step (b)

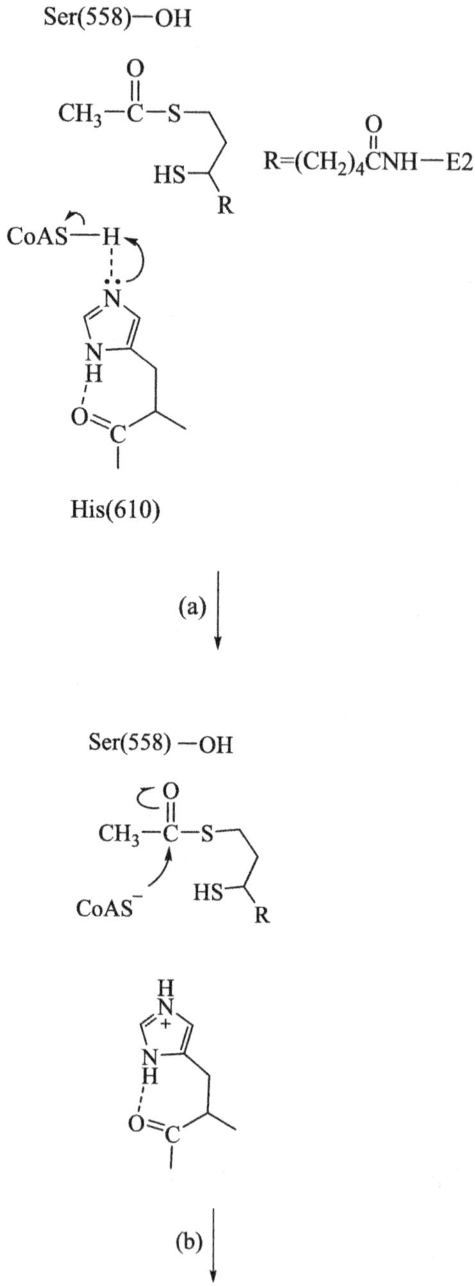

FIGURE 15.6 Detailed mechanism for the formation of acetyl-CoA by dihydrolipoyl transacetylase. *Adapted with permission from Hendle et al.* Biochemistry ***34****, 4287–4289 (1995); Copyright 1995 American Chemical Society.*

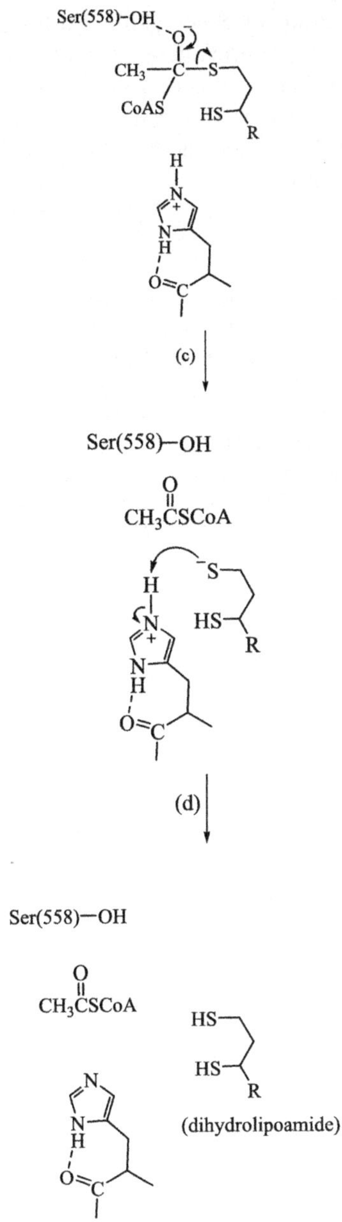

FIGURE 15.6 (continued)

attacks the carbonyl group of S-acetyldihydro-lipoamide-E2p. This leads to the classic tetrahedral intermediate oxyanion, which is stabilized by hydrogen bonding with Ser_{558}. Ejection of the thiolate anion in step (c) is followed by abstraction of a proton from the His_{610} in step (d) to give the enzyme-bound dihydrolipoamide, acetyl-CoA, and the active site amino acids in their original form.

Leading references

Ramachandaren, R. and Schaefer, B. *Chem Texts* **5**, 18 (2019); Patel, M. S.; Nemeria, N. S.; Furey, W.; Jordan, F. *J. Biol. Chem.* **289**, 16615–16623 (2014); Wang, J.; Nemeria, N. S.; Krishnamoorthy, C.; Sowmini, K.; Arjunan, P.; Reynolds, S.; Calero, G.; Brukh, R.; Kakalis, L. O.; Furey, W.; Jordan, F. *J. Biol. Chem.* **289**, 15215–15230 (2014); Yu, X.; Hiromasa, Y.; Tsen, H.; Stoops, J. K.; Roche, T. E.; Zhou, Z. H. *Structure* **16**, 104–114 (2008); Hendle, J.; Mattevi, A.; Westphal, A. H.; Spee, J.; de Kok, A.; Teplyakov, A.; Hol, W. G. J. *Biochemistry* **34**, 4287–4298 (1995); Mattevi, A.; Obmolova, G.; Kalk, K. H.; Teplayakov, A.; Hol, W. G. J. *Biochemistry* **32**, 3887–3901 (1993); Mattevi, A.; Obmolova, G.; Schulze, E.; Kalk, K. H.; Westphal, A.; de Kok, A.; Hol, W. G. J. *Science* **255**, 1544–1550 (1992).

Chapter 16

Farnesyl pyrophosphate synthase

16.1 Farnesyl pyrophosphate synthase

Farnesyl pyrophosphate synthase (EC 2.5.1.10; FPPS; FPP synthase, FPP synthetase, farnesyl diphosphate synthase) is a member of a super family of enzymes (isoprenyl pyrophosphate synthases, prenyl transferases) involved in the consecutive condensation of isoprenoids and polyisoprenoids to ultimately form a wide variety of linear and polycyclic products. Examples of the latter include terpenes [hydrocarbons made up of repeating 5-carbon isopentenyl (isoprene) units], steroids, carotenoids, and retinoids. Enzymes in this family also attach prenyl fragments to (i.e., prenylate) proteins. Subcategories of the family have been defined as (1) "isoprenyl pyrophosphate synthases" (IPPs) that catalyze the formation of linear terpenes by sequential condensation of isopentenyl pyrophosphate with an allylic pyrophosphate, (2) protein prenyl transferases (see above), and (3) prenyl transferases that cause cyclization chemistry.

16.2 Physiological function

FPPS catalyzes the head-to-tail condensation of isopentenyl pyrophosphate (IPP) and dimethylallyl pyrophosphate (DMAPP) to yield geranyl pyrophosphate (GPP). The enzyme then catalyzes the condensation of the latter compound with another molecule of IPP to yield a C_{15} "sesquiterpene": 2E,6E-farnesylpyrophosphate (FPP; see Fig. 16.1). As noted above, these are early-stage reactions in, for example, the biosynthesis of cholesterol, heme A, and ubiquinone. IPP and DMAPP are formed via the "mevalonate pathway," a pathway ubiquitous in all mammals that originates with the condensation of two acetyl-CoA molecules to form acetoacetyl-CoA. Significant structural data have been obtained with avian and human FPPS, as well as with FPPS isolated from *Escherichia coli*.

Enzyme Active Sites and their Reaction Mechanisms. DOI: https://doi.org/10.1016/B978-0-12-821067-3.00016-7

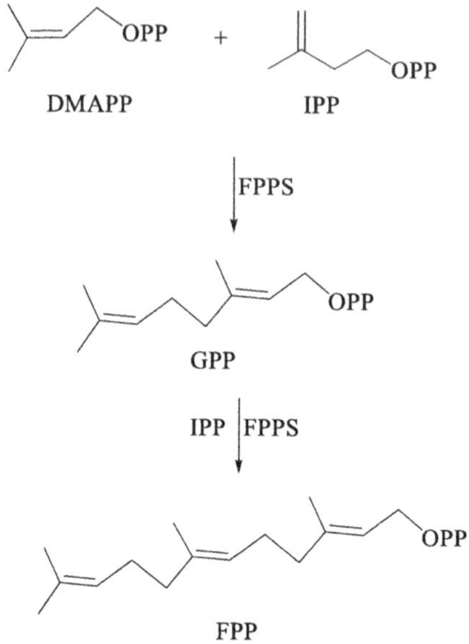

FIGURE 16.1 The farnesyl pyrophosphate synthase condensation reaction.

16.3 Key structural features

FPPS is a homodimer with each subunit containing a single active site. The binding cavity sits at the bottom of a deep cleft, with a highly conserved, aspartate-rich (DDXXD) sequence on each side of the upper (hydrophilic) portion of the cleft. The enzyme requires Mg^{2+} (or Mn^{2+}) for activity, with a cluster of three Mg^{2+} having been shown to position the substrates for bonding (by coordinating to their phosphates groups) and to facilitate the initial ionization step (see below). There are two binding sites within the active site, one for the allylic (DMAPP; GPP) and one for the homoallylic (IPP) substrates. The two DDXXD regions and the trinuclear metal cluster lie within the allylic binding site. A pair of phenylalanine residues is part of a hydrophobic region at the bottom of the allylic substrate-binding pocket that sets the limit on the size of the chain length of the condensation reaction. The residues making up the active site for the *E. coli* and human enzyme are highly conserved.

The human FPPS has three conformational states, "open," "partially closed," and "fully closed." In the reaction of IPP with DMMAP, the DMAPP and metal ions bind first, with the DMAPP bound into the "allylic binding site." This brings the two DDXXD regions closer to one another and creates the "partially closed" conformation. Binding of the IPP into the "IPP binding site" closes this site and generates the "fully closed" conformation. The two substrates are now oriented for reaction with one another and a carbocation intermediate formed during the reaction is shielded from external water.

The IPP and DMAPP binding sites are shown below in Figs. 16.2 and 16.3. The IPP site figure is adapted from an X-ray structure of human FPPS with an FPPS inhibitor, zoledronate, occupying the DMAPP allylic binding site. The DMAPP site figure is adapted from an X-ray study of FPPS in *E. coli* containing dimethylallyl S-thiolodiphosphate (DMSPP) in this site in place of DMAPP. DMSPP is a noncleavable DMAPP analog that contains sulfur in place of oxygen as the bridging (PP to aliphatic chain) atom. A similar study for the human FPPS DMAPP binding site, with the inhibitor, risedronate, occupying this site, are Asp_{103}, Asp_{107}, Arg_{112}, Lys_{200}, Thr_{201}, Asp_{243}, and Lys_{257} (not shown).

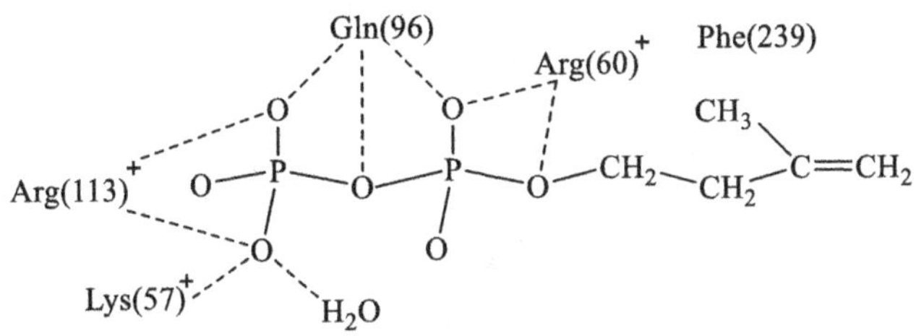

FIGURE 16.2 The IPP binding site. *IPP*, Isopentenyl pyrophosphate. *Adapted with permission from Rondeau et al.* ChemMedChem *1, 267–273 (2006); Copyright 2006 John Wiley and Sons.*

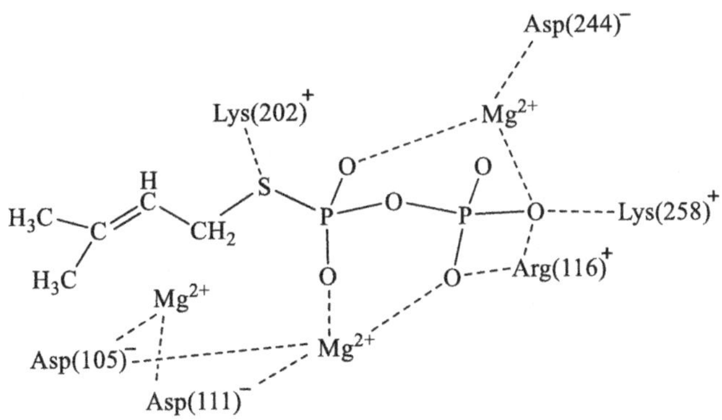

FIGURE 16.3 The DMAPP binding site. *DMAPP*, Dimethylallyl pyrophosphate. *Adapted from Aaron, J. A.; Christianson, D. W.* Pure Appl. Chem. *82, 1585–1597 (2010).*

16.4 Reaction sequence

The condensation of IPP and DMAPP (and subsequently IPP with GPP) is remarkably stereospecific. Early studies elegantly outlined this chemistry using deuterated and tritiated substrates to map the specific C–H bonds cleaved during the course of the reaction. The stereochemical consequences are shown in Fig. 16.4. Note that when the sequence in Fig. 16.4 was elaborated, a nucleophile (X^-) was thought necessary to initiate the reaction, and bond making was thought to be concerted with bond breaking. This is no longer the case and a more current view of the mechanism is shown in Section 16.5. The stereochemistry is, however, incontrovertible, the key features of which are: (1) electrophilic attack on the *si* face of the IPP by C1 of the allylic substrate, with inversion at this carbon, which sets up (2) the stereospecific loss of H_R in step b (as seen by replacing H_S with a tritium atom) to form an E double bond. Note that there are prenyl transferases that form Z-alkenes; these have been shown to cleave the H_S proton.

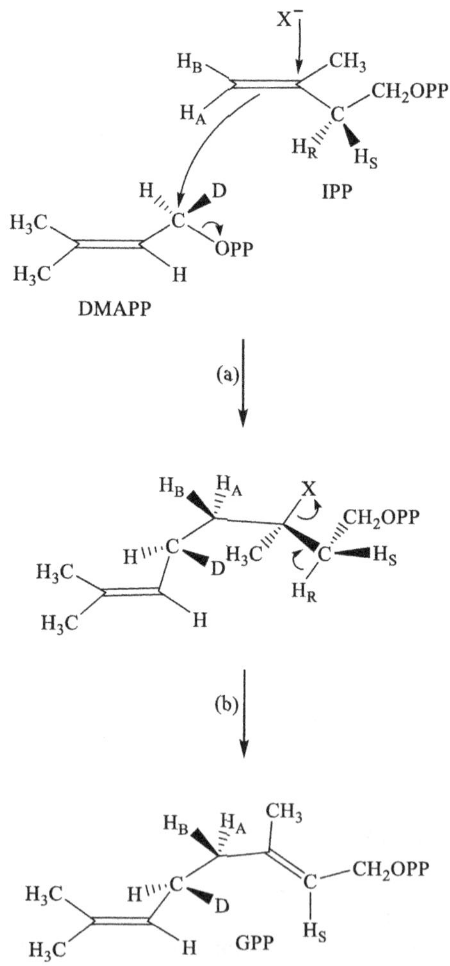

FIGURE 16.4 Stereochemistry of the condensation reaction.

16.5 Detailed mechanism and the role of active site residues

The generally accepted mechanism for the formation of GPP has been termed "ionization–condensation–elimination." The metal cluster forms a complex involving the pyrophosphate groups of both reactants, which (1) facilitates the loss of the DMAPP allylic pyrophosphate leaving group by reducing its negative charge and (2) orients the substrates for C–C bond formation ("condensation"). AA residues that participate in the requisite contiguous placement of the two substrates are among those shown in Figs. 16.2 and 16.3. It is not clear as to whether a discrete carbocation is first formed upon DMAPP cleavage (step a in Fig. 16.5) or that attack by the IPP alkene *si* face is synchronous with the dissociation. In either event, it has been noted that a resonance-stabilized allylic carbocation, if formed, would have a significant barrier to rotation about its C–C bonds, and thus hold its stereochemistry. The pyrophosphate leaving group is stabilized by its complexation with the trinuclear metal cluster, and the developing positive charge by the pyrophosphate, Lys_{202}, Thr_{203}, and Gln_{241}. Condensation (step b in Fig. 16.5) generates a second carbocation, which is then stereospecifically deprotonated by removal of H_R to form the E alkene (step c).

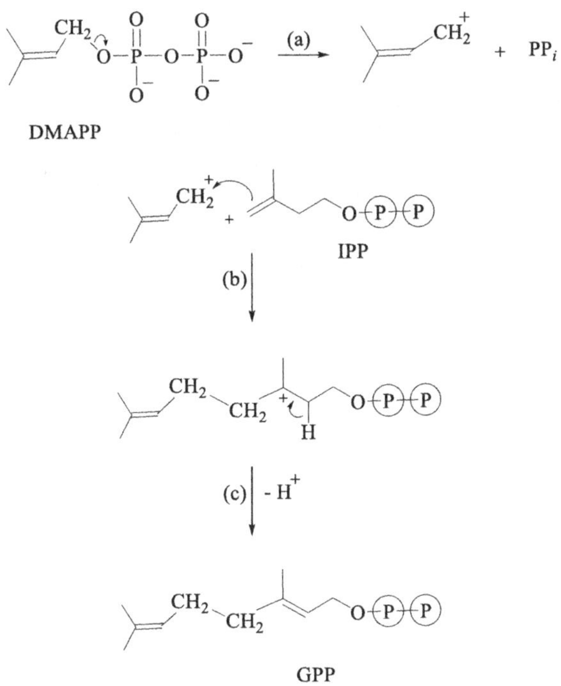

FIGURE 16.5 The ionization–condensation–elimination mechanism for the reaction of DMAPP with IPP to form GPP. *DMAPP*, Dimethylallyl pyrophosphate; *IPP*, isopentenyl pyrophosphate; *GPP*, geranyl pyrophosphate.

An ongoing issue has been identifying the basic species responsible for the proton removal in step c. The magnesium complexed pyrophosphate formed from the DMAPP in step a has been suggested for this role because of its proximity and relative orientation to H_R after the ionization step (see Fig. 16.6).

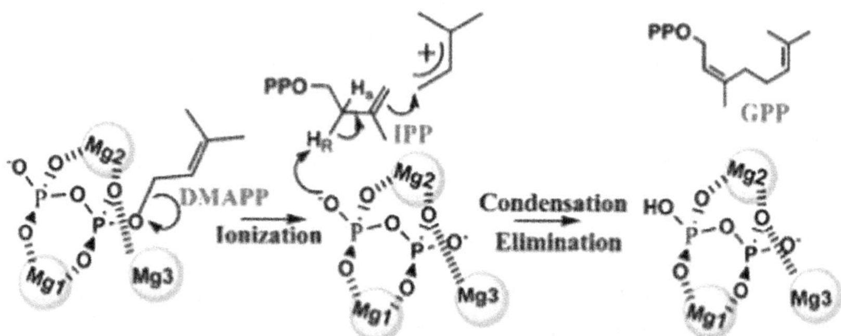

FIGURE 16.6 Magnesium phosphate complex as the base for abstracting H_R. *Republished from the Journal of Biological Chemistry. Hosfield et al.* J. Biol. Chem. *279, 8526–8529 (2004); Copyright 2004 The American Society for Biochemistry and Molecular Biology.*

Leading references

Aaron, H. A. and Christianson, D. W. *Pure Appl. Chem.* **82**, 1585–1597 (2010); Rondeau, J.-M.; Bitsch, F.; Bourgier, E.; Geiser, M.; Hemmig, R.; Kroemer, M.; Lehmann, S.; Ramage, P.; Rieffel, S.; Strauss, A.; Green, J. R.; Jahnke, W. *ChemMedChem* **1**, 267–273 (2006); Kavanagh, K. L.; Guo, K.; Dunford, J. E.; Wu, X.; Knapp, S.; Ebetino, F. H.; Rogers, M. J.; Graham, R.; Russell, G.; Oppermann, U. *Proc. Natl. Acad. Sci. U. S. A.* **103**, 7829–7834 (2006); Hosfield, D. J.; Zhang, Y.; Dougani, D. R.; Broun, A.; Tari, L. W.; Swanson, R. V.; Finn, J. *J. Biol. Chem.* **279**, 8526–8529 (2004); Liang, P.-H.; Ko, T.-P.; Wang, H.-J. *Eur. J. Biochem.* **269**, 3339–3354 (2002); Poulter, C. D. and Rilling, H. C. *Acc. Chem. Res.* **11**, 307–313 (1978); Cornforth, J. W.; Cornforth, R. H.; Popjak, G.; Yengoyan, L. *J. Biol. Chem.* **241**, 3970–3987 (1966).

Chapter 17

Fructose-1,6-bisphosphate aldolase

17.1 Fructose-1,6-bisphosphate aldolase

Fructose-1,6-bisphosphate aldolase (EC 4.1.2.13; aldolase A; FBP aldolase; FBPA) is 1 of 10 enzymes that comprise the glycolytic (Embden–Meyerhof –Parnas) pathway. It is one of three tissue-specific isozymes (A, B, C) found in mammals. Type A are found in muscle and red blood cells, B in liver, kidney, and the small intestine, and C in neuronal tissue. Aldolases are a group of lyase enzymes that catalyze the cleavage of a C–C bond in a β-hydroxyketone by a retro-aldol reaction (see Fig. 17.1). They are divided into two major families—Type I (principally found in eukaryotes), which form Schiff bases between the substrate and an enzyme residue, and Type II (predominant in bacteria and archaea), which utilize a metal ion within their active site. FBP aldolase is a member of the Type I family. The aldolase reaction converts FBP into D-glyceraldehyde 3-phosphate (G3P) and dihydroxyacetone phosphate (DHAP) (see Fig. 17.1: Note that the acyclic structures are Fischer projections). The chemistry is highly stereospecific and reversible; the same enzyme catalyzes C–C bond formation during gluconeogenesis.

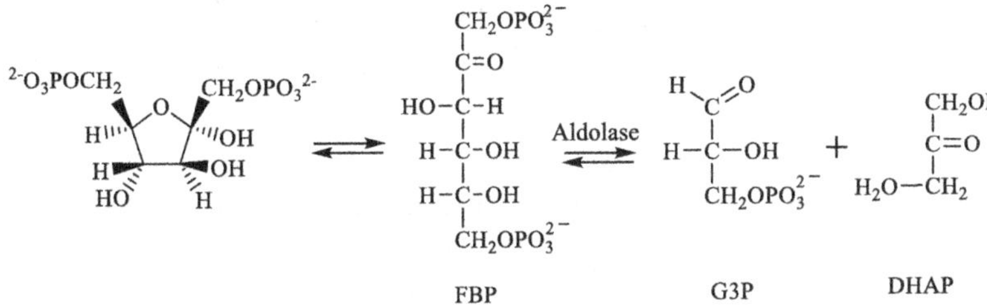

FIGURE 17.1 The overall chemistry catalyzed by FBP aldolase. *FBP*, Fructose-1,6-bisphosphate.

17.2 Physiological function

The metabolism of glucose by the glycolytic pathway converts 1 mole of glucose into 2 moles of pyruvate via FBP, with a net generation of 2 moles

Enzyme Active Sites and their Reaction Mechanisms. DOI: https://doi.org/10.1016/B978-0-12-821067-3.00017-9

of ATP. The sequence is comprised of 10 enzymes, of which aldolase is the fourth. Enolase, found elsewhere in this book, is the ninth enzyme in the glycolytic sequence.

17.3 Key structural features

A key aspect of the cleavage reaction is the requirement to stabilize the anion that is generated when cleavage occurs (see the mechanism below). As noted above, "class I" enzymes convert the ketone into a protonated Schiff base so that cleavage forms a relatively stable "enamine." The "class II" mechanism employs Zn^{2+} complexation to stabilize the enolate anion oxygen. A representation of the Schiff base within the active site of rabbit muscle aldolase is shown in Fig. 17.2. One sees here three of the four residues

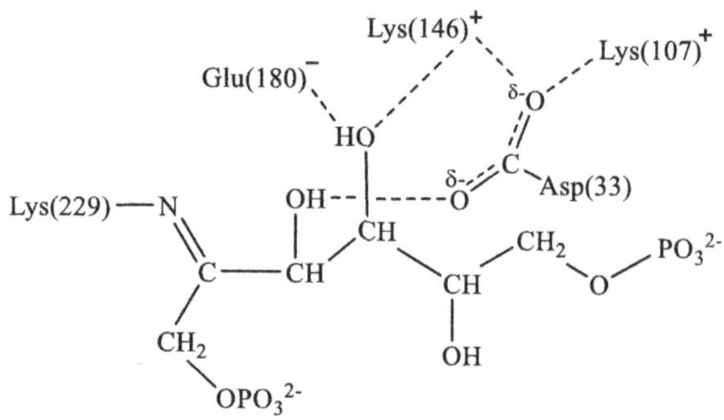

FIGURE 17.2 The active site of rabbit muscle aldolase containing the FBP Schiff base. *FBP*, Fructose-1,6-bisphosphate. *Adapted with permission from Tittmann* Bioorg. Chem. *57, 263–280 (2014).*

that are involved in the catalytic chemistry: Lys_{146}, Glu_{187}, Lys_{229}, and Tyr_{363}. Not shown are the bases that interact with the phosphate groups: P_1: Ser_{271}, Gly_{272}, Gly_{302}, and Arg_{303}; P_6 (in addition to Lys107): Ser_{35} and Ser_{38}. A more extensive image of the human muscle aldolase active site may be found as Fig. 17.2 in Dalby et al. (1999).

17.4 Reaction sequence

The aldol condensation reaction ("condensation" denotes loss of water) is a widely utilized reaction for generating carbon–carbon bonds in synthetic organic chemistry. It is a reversible reaction and cleavage of β-hydroxyketones via the "retro-aldol" reaction is also commonplace. This chemistry can be catalyzed by both acid and base; base catalysis is illustrated in Fig. 17.3.

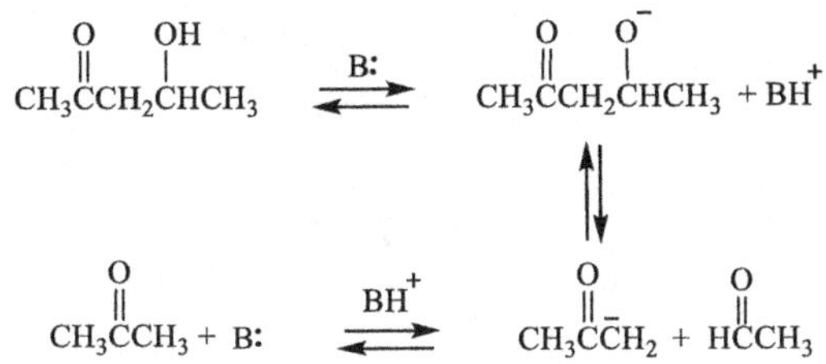

FIGURE 17.3 General mechanism for the base-catalyzed retro-aldol reaction.

Note that cleavage to G3P and DHAP in this manner requires a preceding conversion of glucose (as glucose-6-phosphate) to fructose (as fructose-6-phosphate) to achieve the necessary placement of the β-hydroxycarbonyl element. This is carried out by phosphoglucose isomerase, the second enzyme in glycolysis.

17.5 Detailed mechanism and the role of the active site residues

A mechanism is shown in Fig. 17.4 using AA residue numbering as in rabbit muscle aldolase. As noted above, the key to the catalytic efficiency of aldolase is the initial conversion of FBP into an imine (Schiff base). Lys_{229} provides the amino group and its nucleophilic attack on the ketone is acid-catalyzed by Glu_{187} (step a). Lys_{146} stabilizes the proximal anion, a role it also plays in steps (c) and (e). Not shown is Asp_{33}, which also stabilizes the substrate by H-bonding to the C-3-hydroxyl group and to Lys_{146}. In step (b) the glutamate anion deprotonates the protonated lysine amine, thus setting up the dehydration reaction in step (c). Glu_{187} again provides acid catalysis. The "retro-aldol"-like cleavage of the C–C bond then occurs in step (d), initiated by deprotonation of the hydroxyl group by the glutamate anion. The products are G3P and a resonance hybrid structure that includes an ylid and the neutral enamine). The protonation by Tyr(363) in step (e) is stereospecific, delivering what becomes the H_S hydrogen of the prochiral methylene group. Hydrolysis of the iminium ion releases the DHAP. It should be noted that the active site contains multiple residues capable of providing the protonation/deprotonation sequences seen in this mechanism, a source of some divergence in the literature.

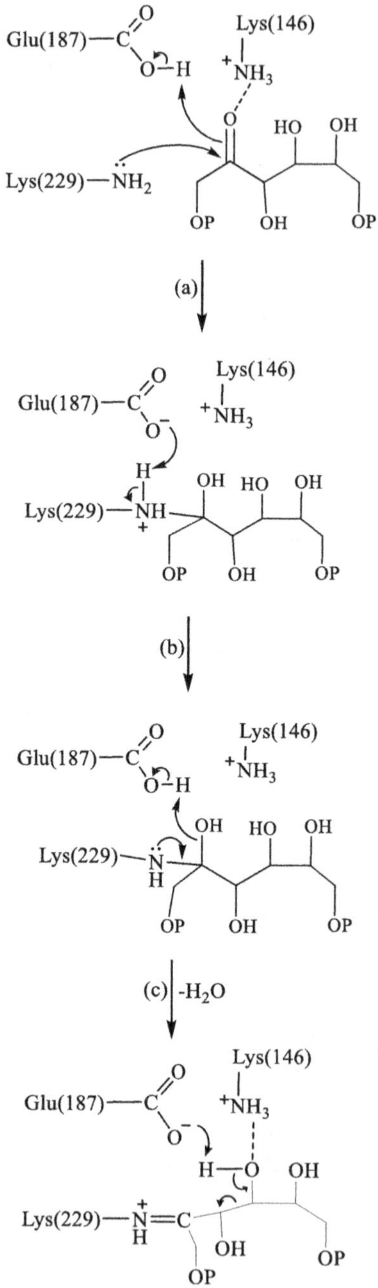

FIGURE 17.4 A detailed mechanism and the role of the active site residues. *From Tittmann* Bioorg. Chem. *57, 263–280 (2014). See Fig. 17.2.*

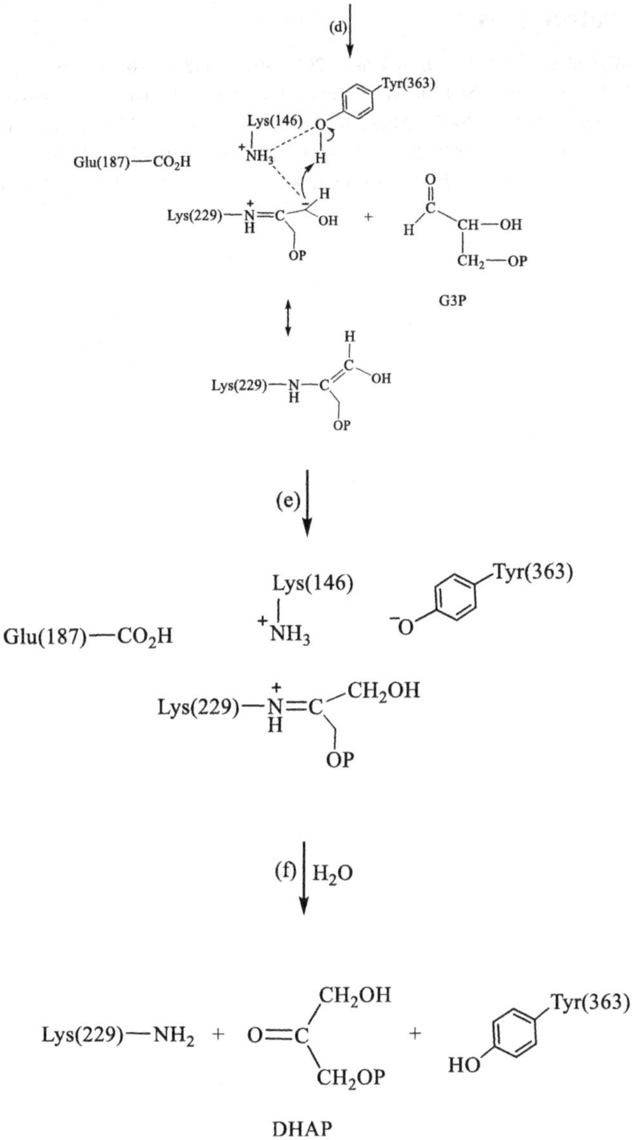

FIGURE 17.4 Continued

Leading references

Hereon, P. W. and Sygusch, J. *J. Biol. Chem.* **292**, 19849–19860 (2017); Tittmann, K. *Bioorg. Chem.* **57**, 263–280 (2014); St-Jean, M.; Lafrance-Vanasse, J.; Liotard, B.; Sygusch, J. *J. Biol. Chem.* **280**, 27262–27270 (2005); Maurady, A.; Zdanov, A.; de Moissac, D.; Beaudry, D. *J. Biol. Chem.* **277**, 9474–9483 (2002); Choi, K. H.; Shi, J.; Hopkins, C. E.; Tolan, D. R.; Allen, K. N. *Biochemistry* **40**, 13868–13875 (2001); Dalby, A.; Dauter, Z.; Littlechild, J. A. *Protein Sci.* **8**, 291–297 (1999).

Chapter 18

Hepatitis C NS2/3 protease

18.1 Hepatitis C NS2/3 protease

Hepatitis C NS2/3 Protease (NS2–3; hepatitis C endopeptidase 2; no EC assignment as of this writing—will be 3.4.22.X) is an essential protein for the infectivity of hepatitis C (HCV). HCV is a retrovirus (a virus that embodies its genetic information as RNA) that causes chronic liver disease. NS2–3 is a "cysteine protease," one of the four major families of protease enzymes (e.g., enzymes that catalyze the hydrolysis of peptide amide bonds; see Fig. 18.1). Other examples within thisclassification include papain and poliovirus 3C protease. Other protease families include serine proteases (see chymotrypsin), aspartic proteases (see HIV-1 protease), and zinc metalloproteases (see carboxypeptidase A).

$$R-\overset{\overset{\large O}{\|}}{C}-NHR' \xrightarrow[\text{Catalyst}]{H_2O} RCO_2H + H_2NR'$$

FIGURE 18.1 Overall chemistry catalyzed by hepatitis C NS2/3 protease.

18.2 Physiological function

HCV is a single-stranded RNA virus, the genome of which encodes a 3011 AA polyprotein that is cleaved into 10 viral proteins. NS2 and NS3 (NS = nonstructural) are among these, and NS2–3 protease is comprised of residues contributed by both. Its function is to cleave the polyprotein at the junction between NS2 and NS3 (between residues 1026 and 1027). It is essential for viral reproduction.

18.3 Key structural features

A crystal structure of the truncated, postcleavage form of NS2/3 protease ($NS2^{PRO}$) shows it to be a dimer, in which two key residues, His_{143} and Glu_{163}, are contributed by one molecular component and a third essential residue, Cys_{184}, by the other. Other key AAs are Pro_{164} and Leu_{217}, which likewise originate from the separate dimer components. The first three residues constitute a "catalytic triad" for the hydrolysis steps (see below). Pro_{164} is thought to provide an optimal structural conformation for, or stabilization of, the dimer, while Leu_{217} is coordinated with all three of the triad AAs of the active site (Fig. 18.2).

Enzyme Active Sites and their Reaction Mechanisms. DOI: https://doi.org/10.1016/B978-0-12-821067-3.00018-0

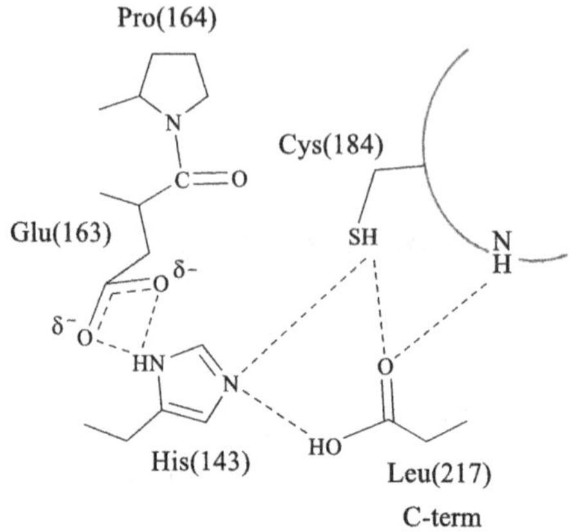

FIGURE 18.2 The active site of NS2[PRO]. *Adapted with permission from Lorenz, I. C.* Viruses **2**, *1635–1646 (2010); by a Creative Commons License, courtesy MDPI.*

The reaction is enhanced by exogenous zinc, which binds tightly to residues in the NS3 portion of the enzyme. It is now accepted that the zinc has no catalytic function, but rather plays a structural role in assuring optimal folding of the enzyme.

18.4 Reaction sequence

This chemistry constitutes a base-catalyzed hydrolysis of an "activated" peptide bond (see Fig. 18.3). Cysteine performs a role equivalent to that which serine plays in the serine proteases.

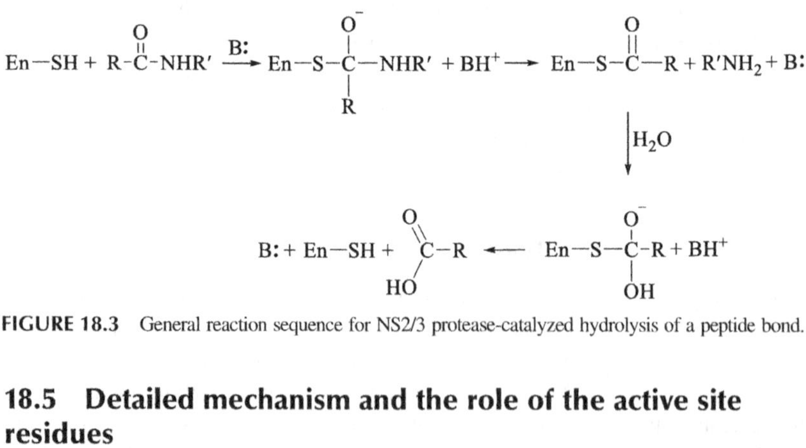

FIGURE 18.3 General reaction sequence for NS2/3 protease-catalyzed hydrolysis of a peptide bond.

18.5 Detailed mechanism and the role of the active site residues

A more detailed description of the mechanism, as it involves the components of the active site, is shown in Fig. 18.4. Though the mechanism for NS2/3 protease

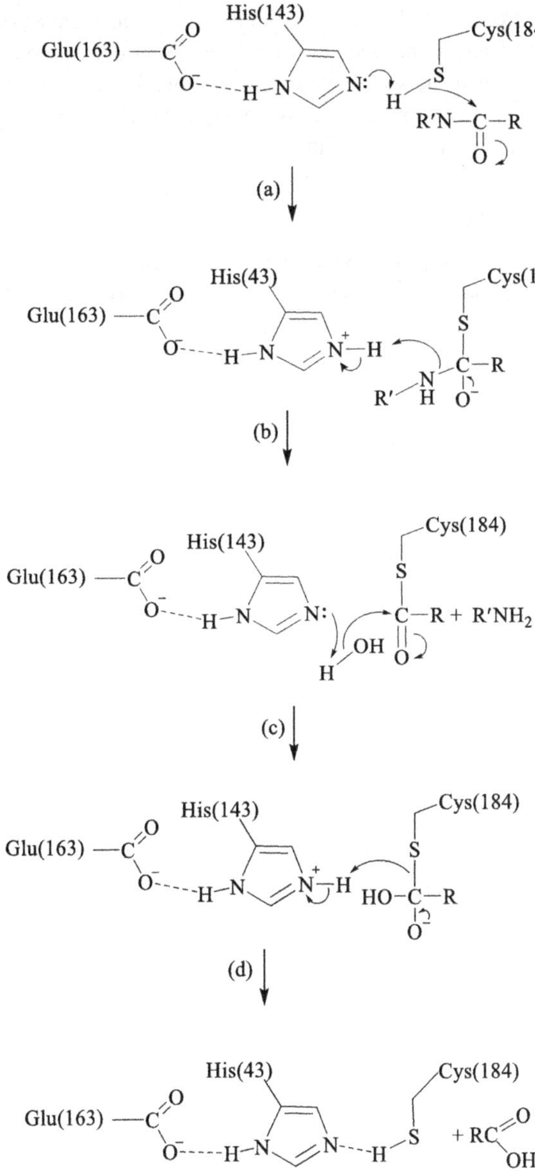

FIGURE 18.4 Detailed mechanism for NS2/3 protease-catalyzed hydrolysis of a peptide bond showing the roles of the key amino acids at the active site.

has not been specifically confirmed, that shown here is by analogy with that of the cysteine protease family. In step (a) His_{143} deprotonates the Cys_{184} thiol group to enable a nucleophilic attack on the amide carbonyl group. The third member of the active site "triad," Glu_{163}, facilitates the deprotonation reaction by forming a hydrogen bond with His_{143}. The amide linkage cleaves in step (b) with the release of an amine and the formation of an enzyme-bound thioester. In step (c) His_{143} again acts as a base in deprotonating a water molecule that attacks the thioester. Cleavage in step (d) releases the Cys_{184} and the carboxylic acid portion of the original peptide linkage.

Leading references

Wu, M.-J.; Shanmugam, S.; Welsch, C.; Yi, M. *J. Virol.* **94**, 1–16 (2020); Lorenz, I. C. *Viruses* **2**, 1635–1646 (2010); Lorenz, I. C.; Marcotrigiano, J.; Dentzer, T. G.; Rice, C. M. *Nature* **442**, 831–835 (2006).

Chapter 19

HIV-1 protease

19.1 HIV-1 protease

HIV-1 protease (EC 3.4.23.16; HIV-1 PR) is required for the infectivity of the retrovirus "HIV" (a retrovirus that embodies its genetic information as RNA). HIV causes "AIDS" (acquired immune deficiency syndrome). PR is an aspartyl protease, one of four major families of protease enzymes (e.g., enzymes that catalyze the hydrolysis of peptide amide bonds). Other enzymes within this classification include pepsins, cathepsins, and renins. Other protease families include serine proteases (see Chapter 9: Chymotrypsin), cysteine proteases (see Chapter 18: Hepatitis C NS2/3 Protease), and zinc metalloproteinases (see Chapter 8: Carboxypeptidase A). The aspartyl and zinc proteases have in common an enhancement of the nucleophilicity of an active site water molecule to initiate its attack on the scissile peptide bond. Aspartyl proteases are also known as acid proteases since these typically have their pH optimum between 3 and 5. The overall reaction catalyzed by PR is shown in Fig. 19.1.

$$R-\overset{\overset{\displaystyle O}{\|}}{C}-NHR' \xrightarrow{H_2O} RCO_2H + H_2NR'$$

FIGURE 19.1 Overall chemistry catalyzed by HIV-1 PR. *HIV-1 PR*, HIV-1 protease.

19.2 Physiological function

There are three enzymes essential to the life cycle of a retrovirus: reverse transcriptase (RT), integrase, and protease. When retroviruses enter a cell its RNA is reverse-transcribed to duplex DNA by RT. The viral DNA enters the cell nucleus and is incorporated into the cellular genetic material by integrase. The host cell transcribes the viral DNA into messenger RNA, which is then translated into viral polyproteins. The role of the protease, HIV PR, is to cleave these polyproteins into structural viral proteins.

Enzyme Active Sites and their Reaction Mechanisms. DOI: https://doi.org/10.1016/B978-0-12-821067-3.00019-2

19.3 Key structural features

PR is a dimer of two identical 99 amino-acid subunits. (Though mammalian aspartases are typically monomeric, the catalytic binding sites of these and their viral analogs are highly conserved as to both their amino acid content and structural topography.) In PR the binding site lies between the dimeric components, with two glycine-rich "flaps" that close down and form hydrogen bonds to the substrate as it binds. The binding site is characterized by two aspartate residues, one from each monomer chain, that are hydrogen-bonded to a single water molecule. There is evidence that one of the aspartate residues is deprotonated and one not.

The Asp(25)-Thr(26)-Gly(27) sequence that characterizes the active site is common to most aspartic proteases. Asp(25) and Asp(25′) are key to the peptide cleavage mechanism. Gly(27) appears to be associated with the release of the N-terminal product. The active site has a pocket to bind aromatic groups adjacent to the scissile bond, with that between Phe and Pro cleaved most efficiently. Protein substrates bind in an extended conformation with approximately 6–7 hydrogen-bound residues at the active site.

A representation of the active site containing a bound polyalanine 6-mer substrate is shown in Fig. 19.2. The key features of this figure can be seen, for example, in the X-ray structure of the diol intermediate (cf. Das et al., 2010).

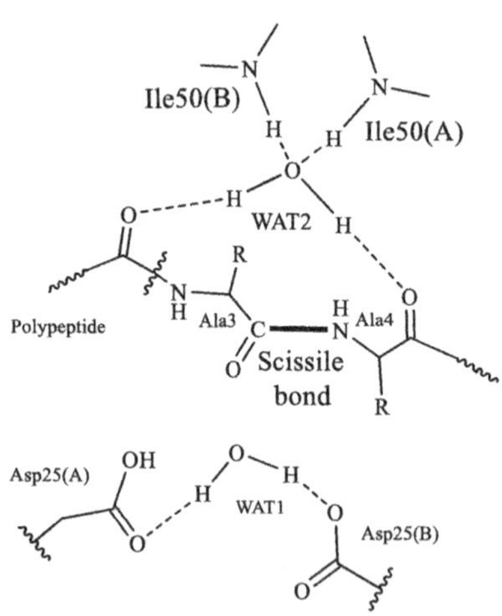

FIGURE 19.2 The computed active site of HIV-1 PR containing a hexameric polyalanine. *HIV-1 PR*, HIV-1 protease. *Adapted with permission from Krzemińska et al.* J. Am. Chem. Soc. *138, 16283–16298 (2016); Copyright 2016 American Chemical Society.*

19.4 Reaction sequence

This chemistry constitutes a base-catalyzed hydrolysis of a peptide bond (see Fig. 19.3). The mechanism differs from, for example, that of the serine proteases in that no covalent bond is formed between the enzyme and the substrate. That alternative was ruled out for PR using $H_2{}^{18}O$.

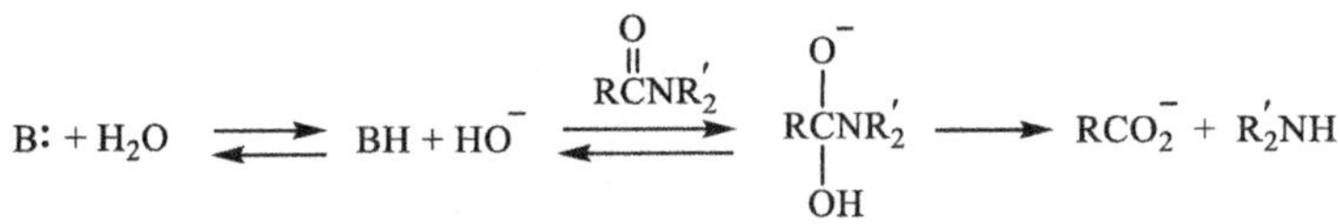

FIGURE 19.3 General reaction sequence for the base-catalyzed hydrolysis of an amide.

19.5 Detailed mechanism and the role of the active site residues

For a step-wise mechanism, there is general agreement that: (1) both active site aspartate residues participate in the reaction, as does an active site water; (2) one of the aspartate (typically, Asp25′) is deprotonated; (3) there is a gem-diol intermediate formed from the carbonyl group involved in the cleavage chemistry. There has been extensive debate in the literature as to whether one of the diol hydroxyl groups is deprotonated—the consensus now appears to be not. Though a concerted mechanism has been ruled out in numerous papers, this possibility has been recently reintroduced as a viable option (cf. Lawal et al., 2019).

A mechanism for the hydrolysis that shows the acid–base roles of the two aspartate residues is shown in Fig. 19.4. In step (a) the removal of a proton from the active site water molecule by an aspartate carboxylate anion initiates the nucleophilic addition of an hydroxyl group to the carbonyl group of the scissile peptide linkage. The addition is facilitated by the simultaneous protonation of the carbonyl oxygen by the other aspartic acid. In step (b) the hydrate intermediate is deprotonated by Asp(25) concomitant with protonation of the amino group by Asp (25′) to form a zwitterion intermediate. Cleavage to the products follows in step (c). Recent computations suggest that steps (b) and (c) may be concerted (cf. Krzemińska et al., 2016).

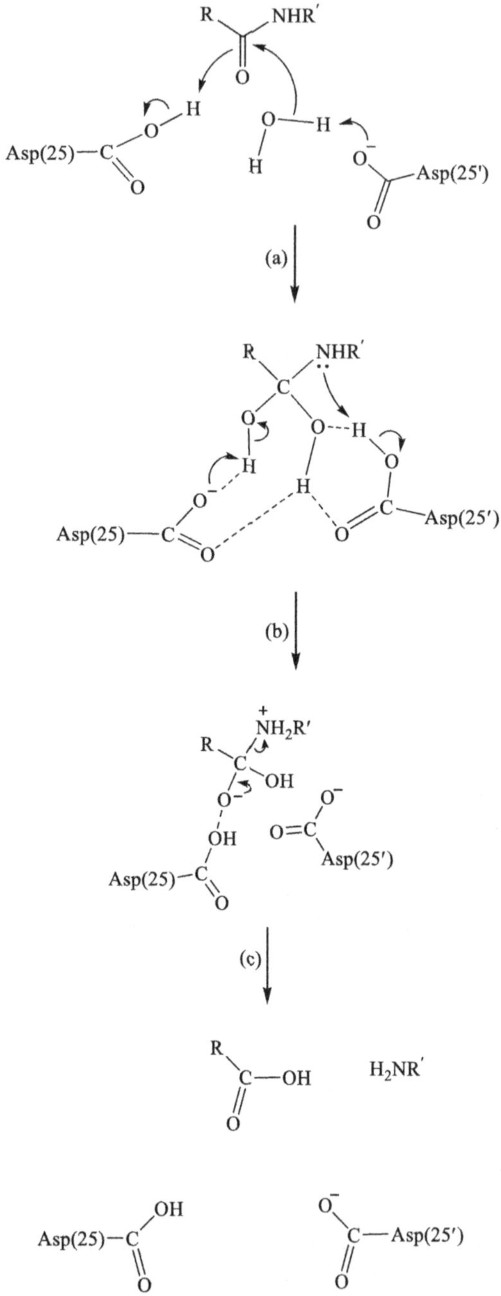

FIGURE 19.4 Detailed mechanism for HIV PR showing the involvement of key amino acids at the active site. *HIV-1 PR*, HIV-1 protease. *From Brik, A. and Wong, C.-H.* Org. Biomol. Chem. *1, 5–14 (2003). Reproduced by permission of the Royal Society of Chemistry.*

Leading references

Lawal, M. M.; Sanusi, Z. K.; Govender, T.; Tolufashe, G. F.; Maguire, G. E. M.; Honarparvar, B.; Kruger, H. G. *Struct. Chem.* **30**, 409–417 (2019); Krzemińska, A.; Moliner, V.; Świderek, K. *J. Am. Chem. Soc.* **138**, 16283–16298 (2016); Shen, C.-H.; Tie, Y.; Wang, Y.-F.; Kovalevsky, A. Y.; Harrison, R. W.; Weber, I. T. *Biochemistry* **51**, 7726–7732 (2012); Kipp, D. R.; Hirschl, J. S.; Wakata, A.; Goldstein, H.; Schramm, V. L. *Proc. Natl. Acad. Sci. U. S. A.* **109**, 6543–6548 (2012); Das, A.; Mahale, S.; Prashar, V.; Bihani, S.; Ferrer, J.-L.; Hosur, M. V. *J. Am. Chem. Soc.* **132**, 6366–6373 (2010); Brik, A. and Wong, C.-H. *Org. Biomol. Chem.* **1**, 5–14 (2003).

Chapter 20

Indoleamine 2,3-dioxygenase-1*

20.1 Indoleamine 2,3-dioxygenase-1

Indoleamine 2,3-dioxygenase-1 (EC 1.13.11.52; IDO; IDO1) is an oxidoreductase dioxygenase that catalyzes the conversion of tryptophan (Trp) to N-formylkynurenine in the first step of the kynurenine pathway. This conversion is shown in Fig. 20.1.

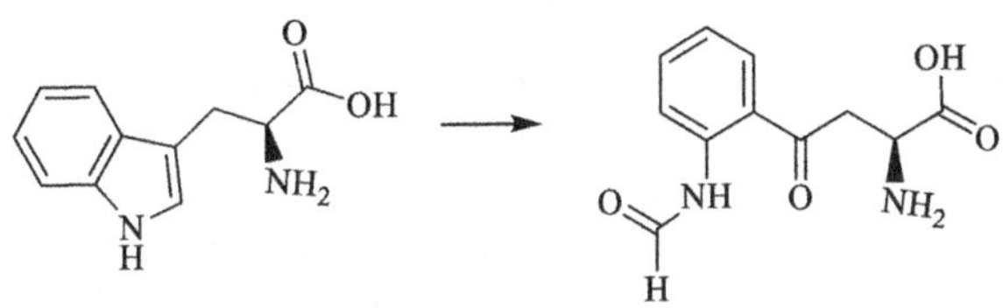

L-Tryptophan N-Formylkynurenine

FIGURE 20.1 The overall chemistry catalyzed by IDO1. *IDO1*, Indoleamine 2,3-dioxygenase-1.

The N-formylkynurenine is subsequently deformylated by formamidase to L-kynurenine and formate. Oxidoreductases catalyze the transfer of electrons between molecules. Oxygenases are metal-containing enzymes that incorporate one or two atoms of oxygen into a substrate. They exist in two classes: monooxygenases and dioxygenases. Dioxygenases typically incorporate both of the atoms of dioxygen into a single substrate and require an essential cofactor, which for IDO1 is heme b ["heme"; iron (II) protoporphyrin IX; Fig. 20.2]. The heme iron must be in its Fe^{2+} (ferrous) oxidation state. Related enzymes are indoleamine 2,3-dioxygenase-2 and tryptophan 2,3-dioxygenase (TDO; see below). IDO1 also exhibits peroxidase activity.

*. I am grateful to Dr. Tito Sempertegui for suggesting, and contributing to, this chapter.

Enzyme Active Sites and their Reaction Mechanisms. DOI: https://doi.org/10.1016/B978-0-12-821067-3.00020-9

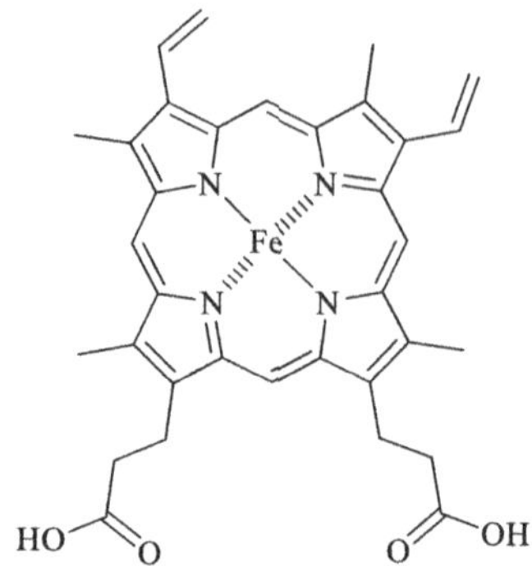

FIGURE 20.2 Iron (II) protoporphyrin IX.

20.2 Physiological function

IDO1 is a monomeric, intracellular enzyme that is ubiquitously expressed in mammalian tissue. It is the rate-limiting enzyme for Trp catabolism outside the liver and, as such, controls the availability of this essential amino acid. Trp is involved in maintaining a strong immune response; its depletion negatively impacts the immune system by reducing the quantity of T-cells. For example, fetal cells initiate Trp metabolism by using IDO1 to suppress T-cell activity so as to tolerate the fetus. As a protein that signals the immune system to ignore a "foreign" substance, IDO1 is termed a "check-point protein." Since there is evidence that tumor cells use IDO1 to inhibit tumor recognition by the immune system, there is a great deal of interest in developing anticancer, "check-point inhibitor" drugs that target this enzyme.

20.3 Key structural features

IDO1 is a monomeric globular enzyme with one large helical domain and one small helical domain connected by a flexible, 17-residue loop that creates a cavity between the domains. The heme lies between these two domains and is connected to the large domain by a histidine residue (His_{346}) on its "proximal" side. The binding of oxygen and L-Trp on the "distal" side leads to the formation of the active ternary complex. A number of amino acids at the active site seem important for orienting the substrate and maintaining the structural integrity of the complex but none have been assigned a specific role in facilitating the chemistry. An X-ray structure is available for the active site of recombinant human IDO (hIDO) complexed with known inhibitors. That for the IDO/4-phenylimidazole (PI) complex is shown in Fig. 20.3. The inhibitor has been removed from the image to make the active site residues more evident.

Of particular note is hydrogen bonding between the heme 6-propionate and Arg_{343} [which itself forms a salt bridge to Asp_{274} (not shown)]. Likewise, it has been suggested that Ser_{263} interacts with the heme 7-propionate. Site-directed mutagenesis studies indicate that Phe_{226}, Phe_{227}, and Ser_{263} are important for activity. This has not been observed for Phe_{163} nor has any specific role been

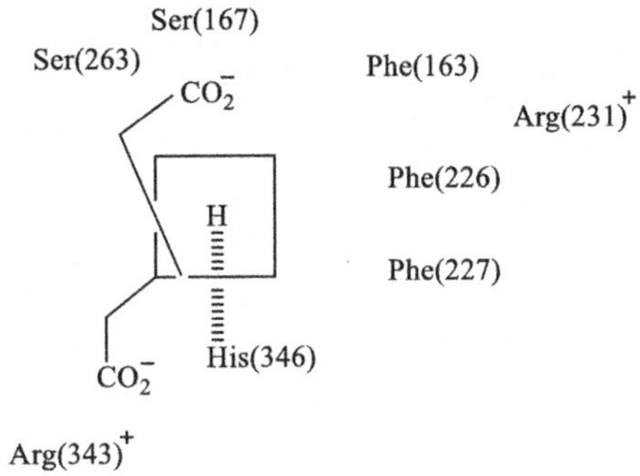

FIGURE 20.3 Schematic rendering of the active site of hIDO containing bound PI (not shown) showing proximal and distal residues. "H" is the heme molecule. *hIDO*, Human IDO; *PI*, IDO/4-phenylimidazole. *Adapted with permission from Sugimoto et al.* Proc. Natl. Acad. Sci. U. S. A. ***103****, 2611–2616 (2006); Copyright 2006 National Academy of Sciences, U.S.A.*

ascribed to Ser_{167}. An X-ray structure of the bacterial enzyme, TDO with Trp bound in the active site, shows very close homology to the structure of IDO with bound PI. Here one finds strong H-bonding between the Tryp carboxylate and an Arg residue equivalent to Arg_{231}, as well as between the Tryp α-amino group and a heme propionate.

20.4 Reaction sequence

The primary intermediate in the oxidative cleavage is the epoxidation of the Trp pyrrole C=C double bond (Fig. 20.4).

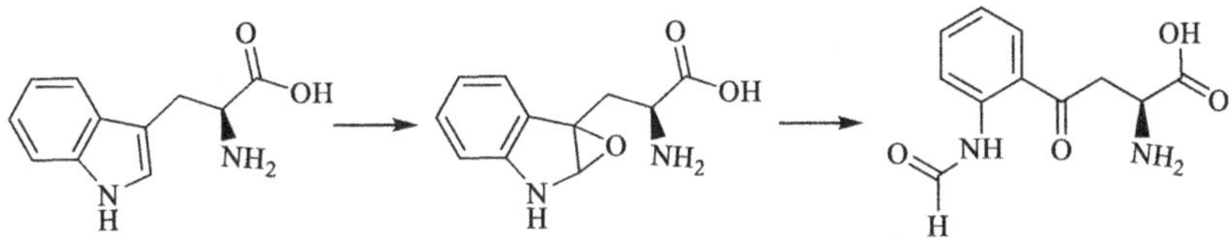

FIGURE 20.4 The IDO1 catalyzed chemistry showing the principal, epoxide, intermediate. *IDO1*, Indoleamine 2,3-dioxygenase-1.

20.5 Detailed mechanism and the role of active-site residues

IDO1 must have the heme iron reduced to ferrous Fe^{II} for O_2 binding. It has been suggested that cytochrome b_5 serves as the activating reductant. The addition of L-Trp to the active site leads to a ternary complex within which the subsequent chemistry occurs. There are few specific details known about the detailed mechanism. Evidence exists for the Trp epoxide and a ferryl species as intermediates, but many other elements of the reaction pathway are conjecture (see also the Chapter 11: Cytochrome P450cam, and Chapter 25: Nonheme Iron Halogenase). Thus it should be recognized that the mechanism presented in Fig. 20.5 and the

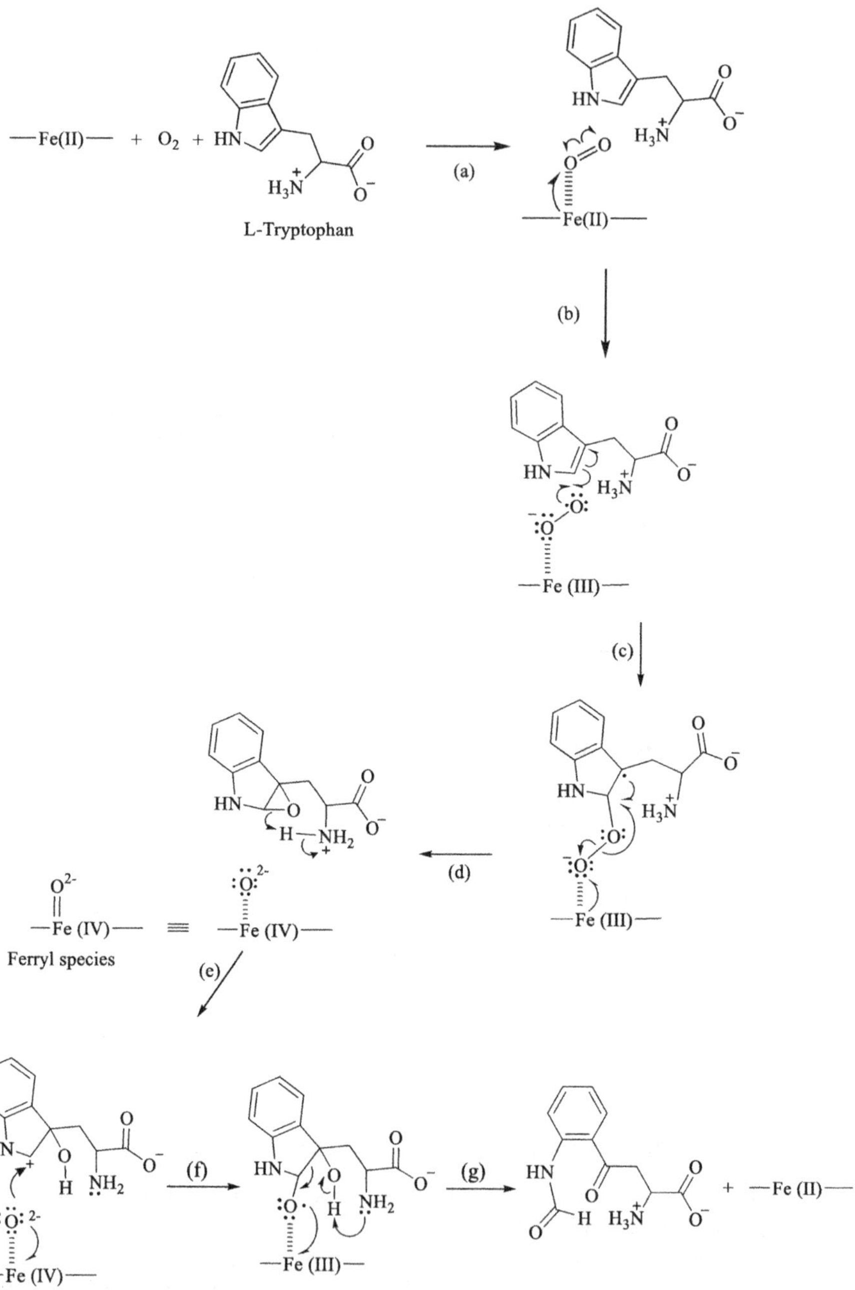

FIGURE 20.5 The mechanism of IDO1 cleavage of L-Trp. *Adapted with permission from Poulos, T. L.* Chem. Rev. ***114****, 3919–3962 (2014); Copyright 2014 American Chemical Society.*

electron-bookkeeping arrow nomenclature within it are speculative. Note that only the heme iron is shown in this mechanism for purposes of simplicity. As will be seen, the active-site AA residues play no direct role in the chemistry but appear to provide structural stability.

The ternary complex is formed in step (a). This is followed by electron transfer from Fe(II) to the dioxygen to form superoxide and the Fe(III) species in step (b). Step (c) involves radical addition of superoxide to the C_2—C_3 double bond of L-Trp, which is followed by (step d) homolysis of the O—O bond to form the epoxide and the ferryl species (note that the spectral signature for this species is consistent with a shortened Fe—O bond, as might be expected for Fe=O). Step (e) is ring opening of the epoxide, assisted by the adjacent amino group, to form a resonance-stabilized cation. Step (f) shows the attack by the ferryl oxygen on the cation (a commonly cited alternative is direct attack on the epoxide) with concomitant formation of Fe(III).The final step (g) is the cleavage of the indole C—C bond to form N-formylkynurenine accompanied by further reduction of the iron to Fe(II).

Leading references

Poulos, T. L. *Chem. Rev.* **114**, 3919–3962 (2014); Freewan, M.; Rees, M. D.; Sempertegui Plaza, T. S.; Glaros, E.; Lim, Y. J.; Wang, X. S.; Yeung, A. W. S.; Witting, P. K.; Terentis, A. C.; Thomas, S. R. *J. Biol. Chem.* **288**, 1548–1567 (2013); Capece, L.; Lewis-Ballester, A.; Yeh, S.-R.; Estrin, D. A.; Marti, M. A. *J. Phys. Chem. B* **116**, 1401–1413 (2012); Efimov, I.; Basran, J.; Thackray, S. J.; Handa, S.; Mowat, C. G.; Raven, E. L. *Biochemistry* **50**, 2717–2724 (2011); Sugimoto, H.; Oda, S.-I.; Otsuki, T.; Hino, T.; Yoshida, T.; Shiro, Y. *Proc. Natl. Acad. Sci. U. S. A.* **103**, 2611–2616 (2006).

Chapter 21

Lysine 2,3-aminomutase

21.1 Lysine 2,3-aminomutase

Lysine 2,3-aminomutase (EC 5.4.3.2; LAM) is one of the best characterized members of the radical-SAM superfamily. LAM converts L-α-lysine into L-β-lysine (L-3,6-diaminohexanoate) as well as the reverse reaction (see Fig. 21.1). The radical-SAM enzymes are noteworthy for their ability to carry out chemistry involving typically unreactive C—H bonds. They share in common the generation of the 5′-deoxyadenosyl 5′-yl radical (5′-dA$^{\bullet}$; dAdo$^{\bullet}$), which plays a critical role in chemistry analogous to that observed with coenzyme B_{12}-dependent enzymes. The source of the dAdo$^{\bullet}$ is the cofactor, *S*-adenosylmethionine (SAM; see Fig. 21.2A). Other requisite cofactors are pyridoxal phosphate (PLP; see Fig. 21.2B) and a $[4Fe-4S]^{2+}$ iron cluster. Radical-SAM enzymes are divided into those, such as LAM, that use SAM catalytically, and those that consume SAM during the course of their chemistry.

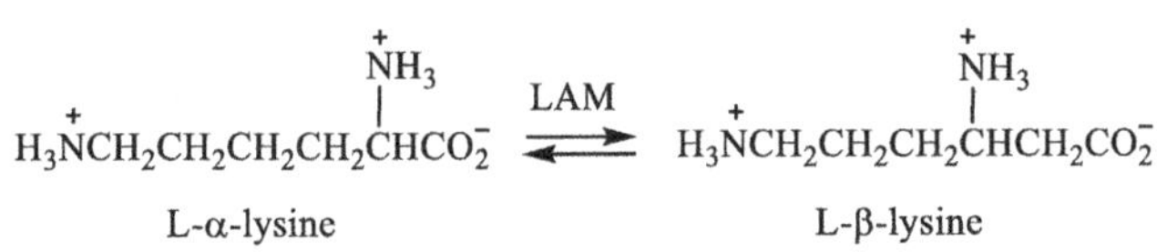

FIGURE 21.1 The overall chemistry catalyzed by lysine 2,3-aminomutase.

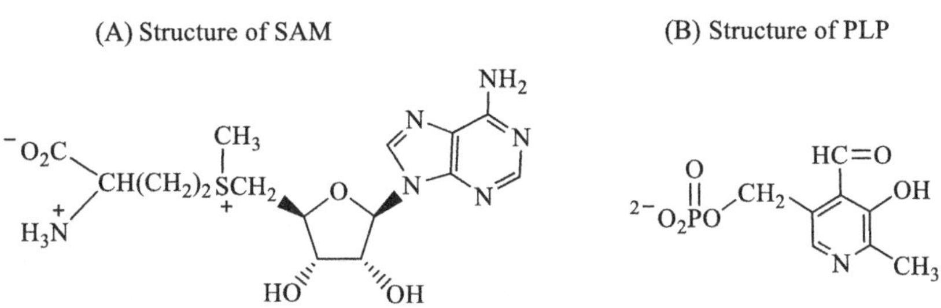

FIGURE 21.2 Structures of (A) SAM and (B) PLP. *SAM*, *S*-adenosylmethionine; *PLP*, pyridoxal phosphate.

Enzyme Active Sites and their Reaction Mechanisms. DOI: https://doi.org/10.1016/B978-0-12-821067-3.00021-0

21.2 Physiological function

LAM has been isolated and characterized from the bacterium *Clostridium subterminale* strain SB4 (*C.* SB4) where it initiates the catabolic conversion of L-α-lysine to acetyl-CoA and ammonia as the bacterium's sole source of carbon and nitrogen. The enzyme is readily degraded by air. It has also been isolated from *Bacillus subtilis*; in this case, it is stable in air but also considerably less active than the clostridial enzyme. Though β-amino acids are not commonplace in nature, β-lysine is used by microbes for the biosynthesis of several members of the streptothricin family of antibiotics.

21.3 Key structural features

The enzyme has been isolated from *C.* SB4 where the active site is located deep within the enzyme, well-removed from the protein surface. This is one of the structural motifs thought to be important for controlling and optimizing the highly reactive free radicals that are generated during the transformation. Three iron atoms of the cluster are each bound to a cysteine residue (Cys_{125}, Cys_{129}, and Cys_{132}); SAM is chelated to the fourth (unique) iron through its amino and carboxylate groups (Fig. 21.3).

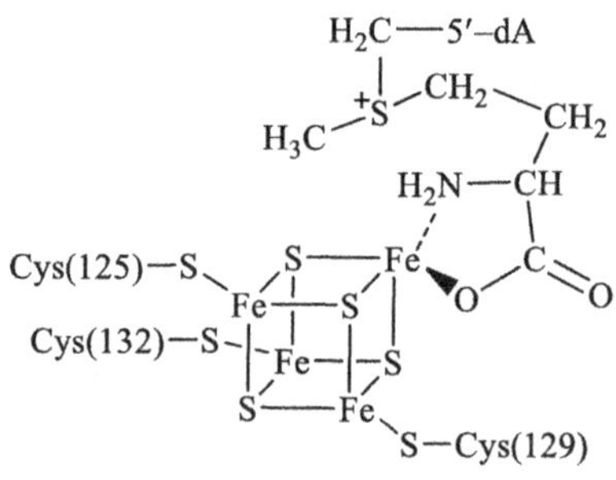

FIGURE 21.3 SAM chelated to the Fe–S cluster. *SAM, S*-adenosylmethionine. *Adapted with permission from Wang, S. C. and Frey, P. A.* Biochemistry *46, 12889–12895 (2007). Copyright 2007, American Chemical Society.*

The close approach of the SAM sulfur atom to the unique iron is important to the electron-transfer step that occurs between them and initiates the SAM cleavage chemistry (see below). A more detailed rendering of the cluster/SAM complex, with key associated AA residues, is presented in Fig. 21.4 (note that a selenium analog of SAM was used for this study). A protein-bound water molecule, held in place by Gln (258), that H-bonds to both a ribose hydroxyl and the SAM carboxylate carbonyl groups and a "glycine-rich" region that interacts with the SAM α-amino group bonded to the cluster are of particular note.

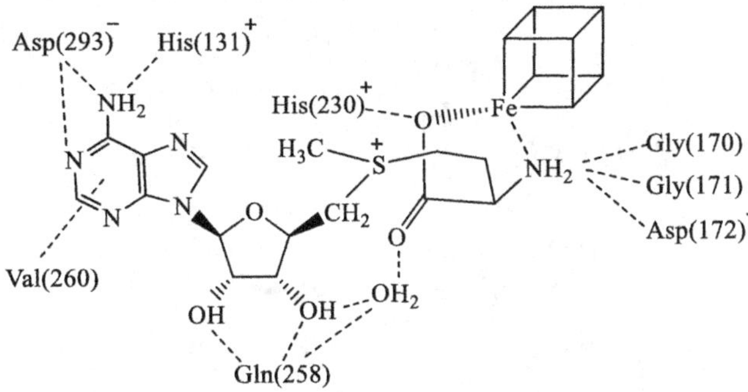

FIGURE 21.4 The SAM/cluster complex showing key AA residues. *SAM*, *S*-adenosylmethionine; AA, amino acid. *Adapted with permission from Vey, J. L. and Drennan, C. L.* Chem. Rev. *111, 2487–2506 (2011). Copyright 2011, American Chemical Society.*

The mutase chemistry involves the initial formation of an aldimine between the lysine substrate and PLP. This aldimine lies adjacent to the 5′-dA portion of the SAM molecule (as required for the chemical reactions involving the 5′-dA• and the lysine C − H bond that characterize the mutase mechanism; see below). Fig. 21.5 shows the aldimine and key AA residues with which it interacts. The interaction of an enzyme-bound water with the PLP pyridine nitrogen atom has been taken as an indication that this nitrogen is unprotonated. Note the nearby Lys(337); as will be seen, this residue initially binds the PLP to the active site as an aldimine. A rendering of the active site containing bound aldimine, a selenium analog of SAM, and the iron cluster is shown in Fig. 21.6.

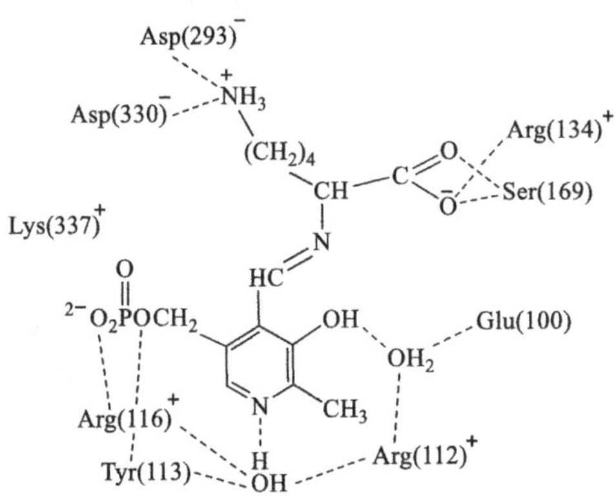

FIGURE 21.5 PLP/lysine aldimine AA interactions at the active site. See Fig. 21.4. *PLP*, Pyridoxal phosphate; *AA*, amino acid. *Copyright 2011, American Chemical Society.*

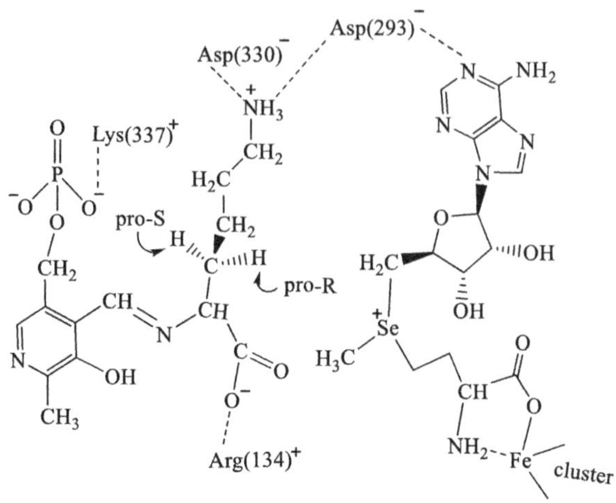

FIGURE 21.6 The LAM active site containing a seleno analog of SAM, the substrate aldimine and the iron cluster. *LAM*, Lysine 2,3-aminomutase; SAM, *S*-adenosylmethionine. *Adapted from Lepore B. W. et al.,* Proc. Natl. Acad. Sci. *102, 13819–13824 (2005). Copyright 2005 National Academy of Sciences, U.S.A.*

21.4 Reaction sequence

The (reversible) reaction sequence is shown in Fig. 21.7.

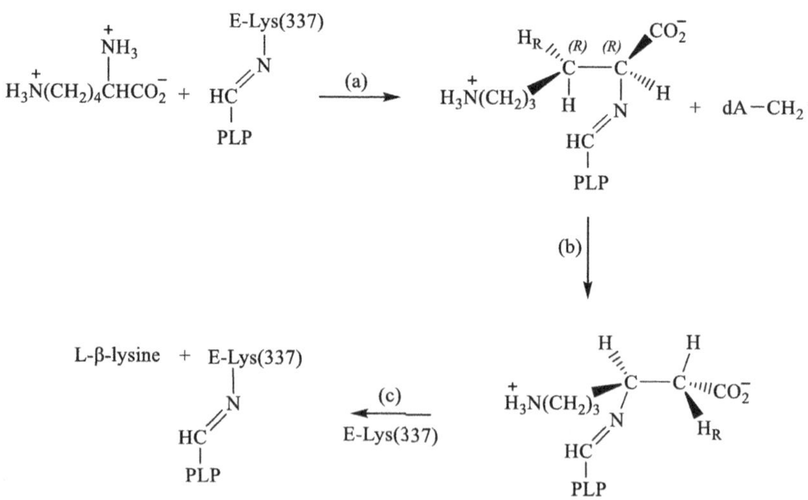

FIGURE 21.7 The reaction sequence catalyzed by LAM. *LAM*, Lysine 2,3-aminomutase.

21.5 Detailed mechanism and the role of active site residues

The mechanism for the isomerization chemistry is shown in Fig. 21.8. There are three initial steps required before the LAM chemistry can take place. In step (a), α-lysine is bound into the active site, where it interchanges with Lys_{337} as an aldimine derivative of PLP. It is shown here as the initial step since it has been suggested that step (b) would likely occur only after the substrate is bound, so as to make the active site ready for the subsequent radical chemistry. It is also known that the presence of substrate within the active site greatly enhances the rate of the cleavage chemistry. Step (b) shows the one electron reduction of the 2+ iron cluster by flavodoxin or

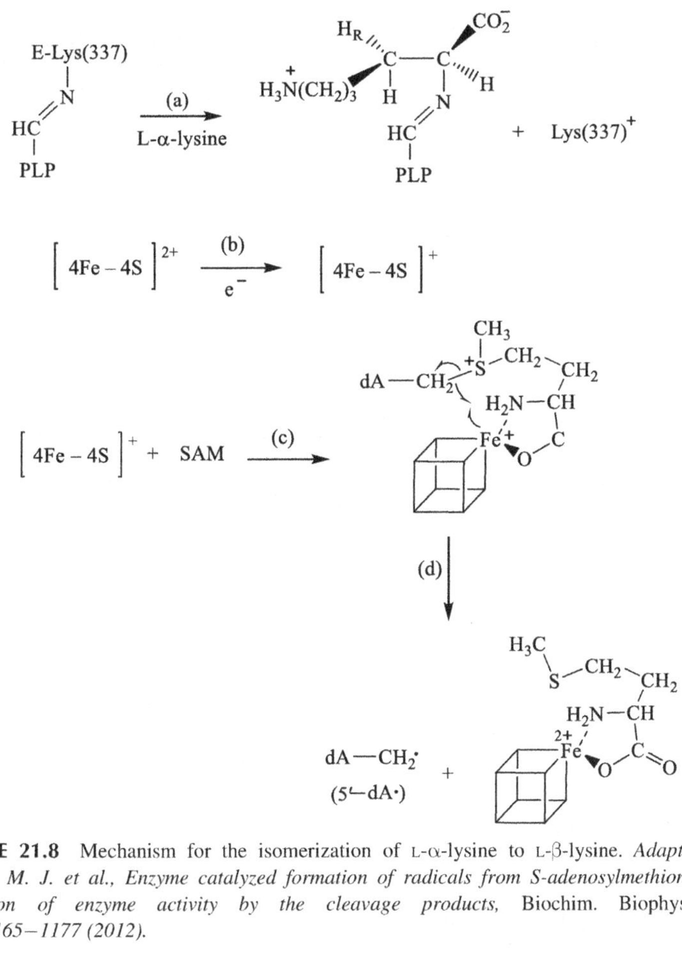

FIGURE 21.8 Mechanism for the isomerization of L-α-lysine to L-β-lysine. *Adapted from Hiscox, M. J. et al., Enzyme catalyzed formation of radicals from S-adenosylmethionine and inhibition of enzyme activity by the cleavage products,* Biochim. Biophys. Acta *1824,1165–1177 (2012).*

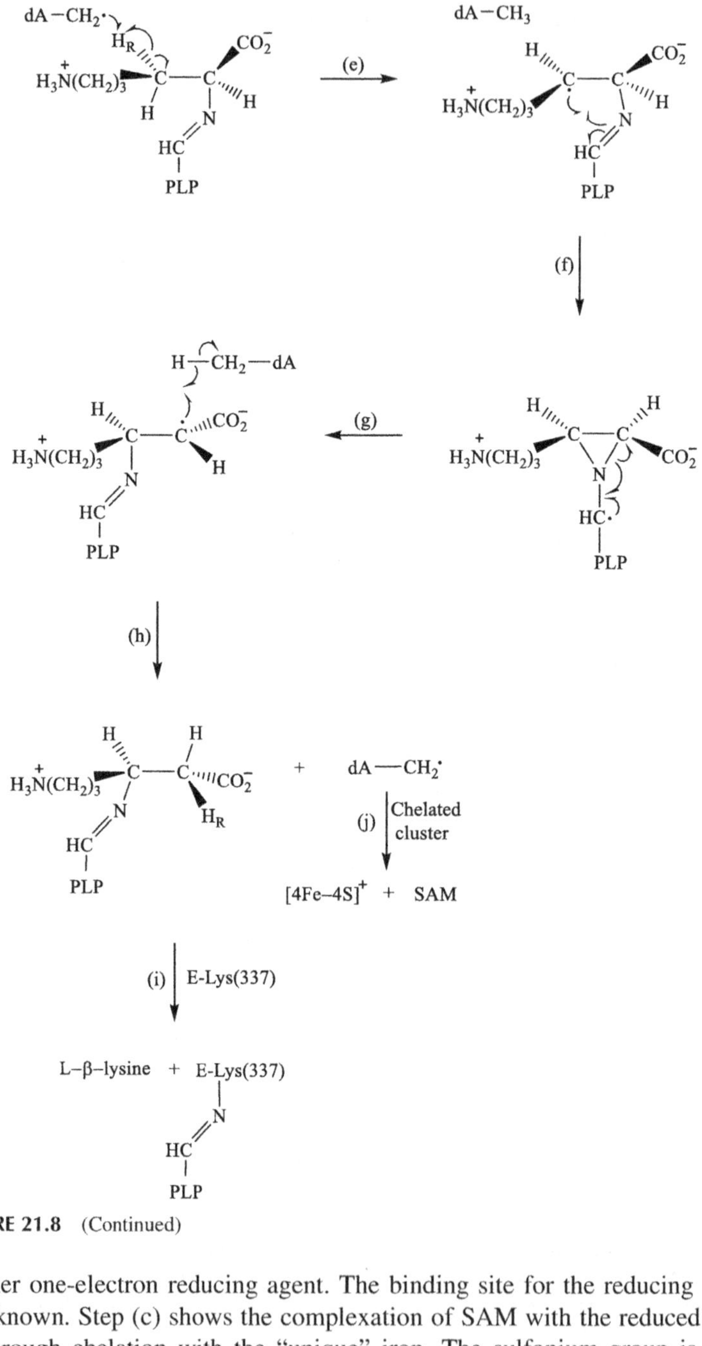

FIGURE 21.8 (Continued)

another one-electron reducing agent. The binding site for the reducing agent is unknown. Step (c) shows the complexation of SAM with the reduced cluster through chelation with the "unique" iron. The sulfonium group is positioned close enough to the cluster to allow orbital overlap between the (trans) S-(C5′) antibonding orbital and the iron atom. Inner sphere electron transfer (step (d)) results in the reductive homolysis of the dA-CH_2—S bond to form the 5′-DA• and a hexacoordinated, oxidized cluster. The specific cleavage of only the methylene—S bond is attributed to the favorable stereoelctronics. The additional bonding at the unique iron is thought to help drive this chemistry. Though the 5′-dA• has not been directly observed, an unsaturated analog has been used to generate a more stable, detectable allylic radical.

In step (e), the 5′-dA• abstracts a hydrogen atom from the substrate, regiospecifically (at C(α)) and stereospecifically (selectively the 3-pro-*R* atom). The basis for the stereospecificty can be seen in Fig. 21.6. The α-lysine β-methylene carbon is c.3.8 Å from the 5′-deoxyadenosyl methylene group, with the pro-*R* hydrogen pointed directly at the incipient radical site.

Radical addition to the imino double bond (step (f)) forms an azocyclopropylcarbinyl radical that cleaves in step (g) to form a carbon radical at the "alpha" position. In step (h), the hydrogen is returned to the substrate by the dA-CH_3 group, again stereospecifically to the 2-pro-*R* position of the β-lysine/PLP aldimine. Clearly, residues at the active site exert exquisite control of the substrate and 5′-dA• orientations to allow such stereospecificity. The product is released, with reformation of the enzyme bound PLP aldimine, in step (i). SAM and the cluster monocation are regenerated in step (j). Note that SAM thus functions catalytically in LAM.

Leading references

Broderick, J. B.; Duffus, B. R.; Duschene, K. S.; Shepard, E. M. *Chem. Rev.* **114**, 4229–4317 (2014); Hiscox, M. J.; Driesener, R. C.; Roach, P. L. *Biochim. Biophys. Acta* **1824**, 1165–1177 (2012); Vey, J. L. and Drennan, C. L. *Chem. Rev.* **111**, 2487–2506 (2011); Wang, S. C. and Frey, P. A. *Biochemistry* **46**, 12889–12895 (2007); Lees, N. S.; Chen, D.; Walsby, C. J.; Benshad, E.; Frey, P. A.; Hoffman, B. M. *J. Am. Chem. Soc.* **128**, 10145–10154 (2006); Lepore, B. W.; Ruzicka, F. J.; Frey, P. A.; Ringe, D. *Proc. Natl. Acad. Sci.* **102**, 13819–13824 (2005).

Chapter 22

Lysozyme

22.1 Lysozyme

Lysozyme (EC 3.2.1.17) is a hydrolytic glycosidase [(β-) glycoside hydrolase; GH] that is a member of a ubiquitous super family with over 100 subfamilies. Lysozyme is a member of the GH subfamily 22. The most extensively studied lysozyme is isolated from hen egg white (HEWL; a "C-type" lysozyme) and consists of a single polypeptide chain of 129 amino acids. The bacteriophage T4 lysozyme has also been widely studied. The net chemistry elicited by this enzyme is an acid-catalyzed hydrolysis of an acetal to a hemiacetal (Fig. 22.1).

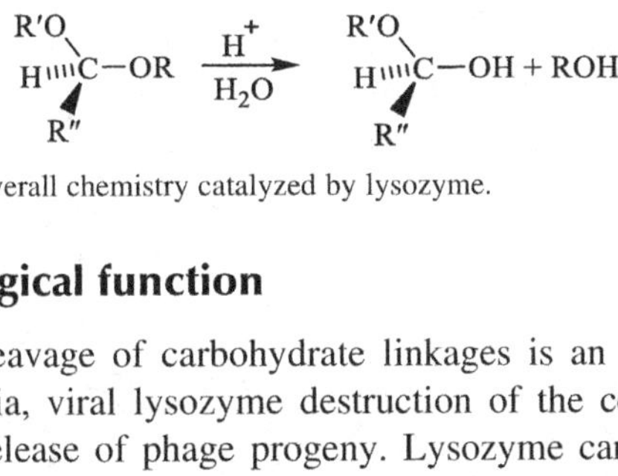

FIGURE 22.1 The overall chemistry catalyzed by lysozyme.

22.2 Physiological function

The hydrolytic cleavage of carbohydrate linkages is an essential metabolic process. In bacteria, viral lysozyme destruction of the cell wall is required for the ultimate release of phage progeny. Lysozyme carries out the hydrolytic cleavage of a β-1,4 glycosidic linkage (α-glycosidases are also known) between the repeating, alternating units of N-acetylmuramic acid (NAM) and N-acetylglucosamine (NAG) that make up one of the polysaccharide components of cell wall peptidoglycans. The site of cleavage is shown in Fig. 22.2.

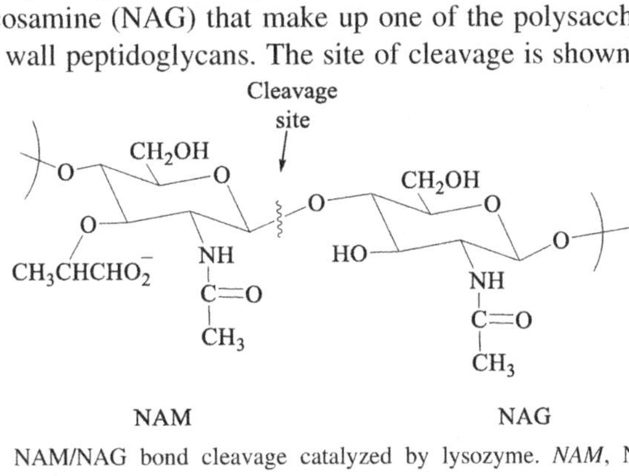

FIGURE 22.2 NAM/NAG bond cleavage catalyzed by lysozyme. *NAM*, N-Acetylmuramic acid; *NAG*, N-acetylglucosamine.

Enzyme Active Sites and their Reaction Mechanisms. DOI: https://doi.org/10.1016/B978-0-12-821067-3.00022-2

22.3 Key structural features

The enzyme folds into two lobes that are connected by an extended helix. The active site resides within a cleft between these lobes. Trp_{63} and Trp_{108} lie on the "southern" and "northern" borders of the cleft. This is readily seen in Fig. 22.3, an X-ray image of HEWL.

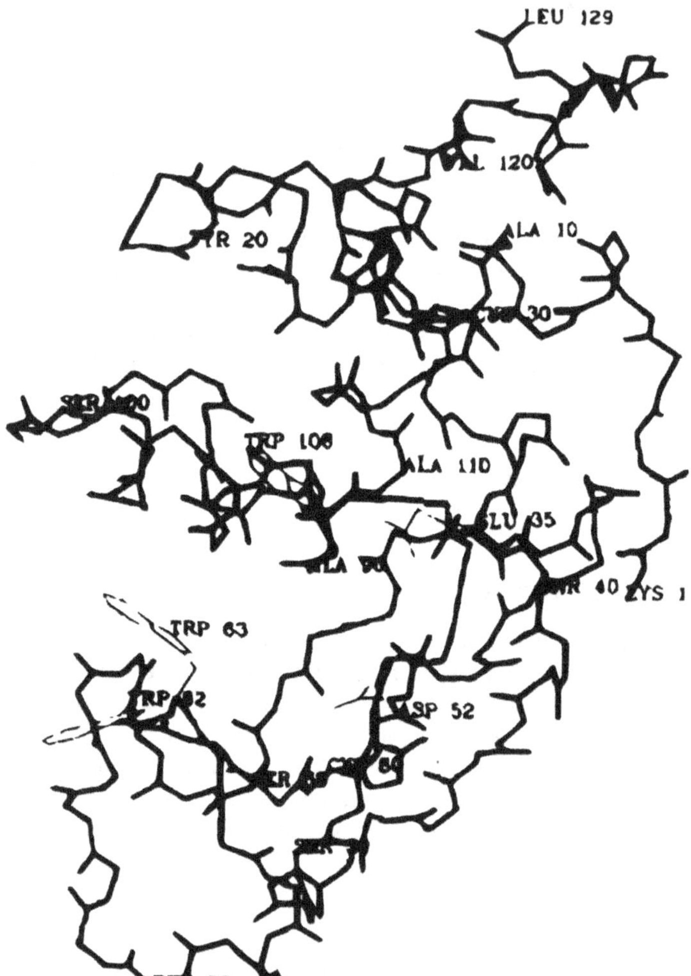

FIGURE 22.3 The structure of HEWL. *HEWL*, Hen egg white lysozyme. *From Strynadka, N. C. J.; James, M. N. G.* J. Mol. Biol. ***220****, 401–424 (1991).*

Modeling led early workers to propose that a hexamer of alternating NAG and NAM residues would fit into the active site, and this was later confirmed for the bacterial cell wall. The substrate lies within a successive set of subsites

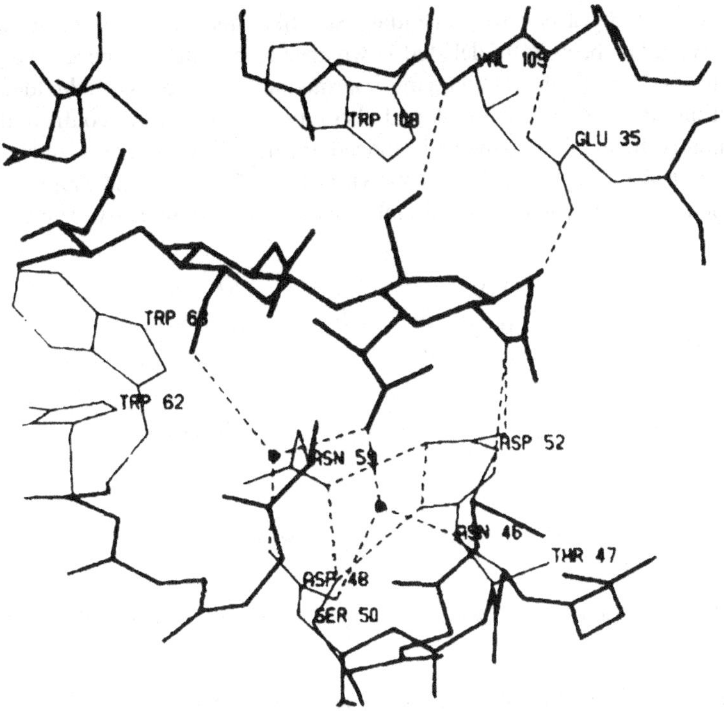

FIGURE 22.4 A NAM residue bound into the subsite D of the active site. Hydrogen bonds are shown as dashed lines. *NAM*, N-Acetylmuramic acid. *From Strynadka, N. C. J.; James, M. N. G.* J. Mol. Biol. ***220**, 401–424 (1991).*

labeled A–F. Cleavage takes place between sites D and E. The X-ray structure of a NAM residue bound into subsite D is shown in Fig. 22.4.

As will be discussed below, the key active site residues in HEWL have been identified as Glu_{35} and ASP_{52} (see Fig. 22.4).

22.4 Reaction sequence

This reaction is an acid-catalyzed hydrolysis of an acetal to a hemiacetal. Such chemistry typically proceeds through a mechanism termed "S_N1 solvolysis," in which the cleavage step involves the expulsion of an alcohol to form an oxocarbenium ion intermediate. Here, that cation would be greatly stabilized by resonance involving the adjacent oxygen. However, such (planar) carbocations are then normally attacked by water to give a mixture of stereoisomeric hydroxyl compounds that, in this case, would result in both retention and inversion at C1. In fact, the chemistry catalyzed by GH s are typically stereospecific, resulting in either inversion or retention at the anomeric carbon.

Lysozyme is a "retaining" GH (there is an alternative class of "inverting" GH). The current view as regards the mechanism favors a "double-inversion" process whereby a pair of sequential "S_N2-like" reactions lead to net retention. What has been hotly debated is whether or not this sequence involves a covalently bound substrate-enzyme intermediate or an enzyme-shielded carbocation. Recent experimental and theoretical results now confirm that a covalently bound intermediate is indeed formed. The general sequence for such hydrolysis with retention is shown in Fig. 22.5 (note that Enz-nu:$^-$ corresponds to a nucleophilic functional group at the enzyme active site).

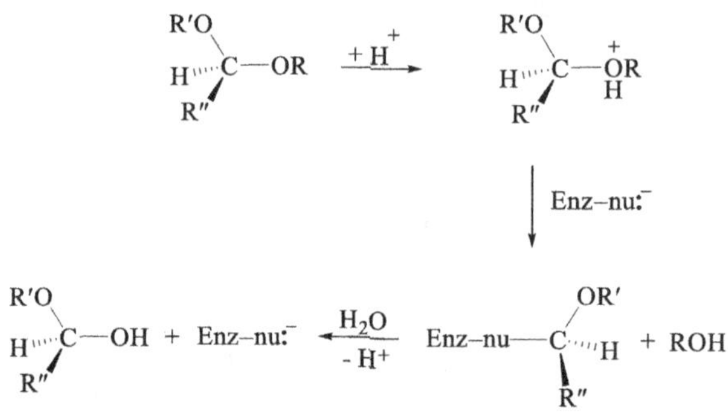

FIGURE 22.5 General reaction sequence for enzymic hydrolysis of an acetal with net retention.

22.5 Detailed mechanism and the role of the active site residues

The key features of the chemistry in Fig. 22.5 are the presence of a proton donor and a nucleophilic residue capable of forming the tetrahedral intermediate. In lysozyme these roles are filled by Glu_{35} and Asp_{52}, respectively. Because Asp_{52} is surrounded by polar groups (Asn_{46} and Ser_{50}) it has a normal carboxylic acid pKa and is unprotonated at physiological pH. In contrast, the acid functional group of Glu_{35} lies in a nonpolar pocket (adjacent are Ala_{110} and Trp_{108}). Its environment inhibits ionization and spectral data confirm that it is not ionized at physiological pH. The mechanism is shown in Fig. 22.6.

Step (a) shows the protonation of the acetal by Glu_{35}. The attack by the nucleophile, Asp_{52}, leads to C-O cleavage at the anomeric site via a transition state with partial positive and negative charges spread over multiple sites [step (b)]. Noteworthy are the stabilization of the developing charge at the anomeric carbon by the adjacent oxygen, and the distortion of the cyclohexyl ring to provide the necessary stereoelectronic arrangement of the participating orbitals and to avoid clashing with Trp_{108}. This leads to the release of NAG-OH and the formation of the covalently bonded enzyme-substrate intermediate with an inversion in step (c). Water within the active site is then deprotonated by the Glu_{35} anion and moves to displace the aspartate [step (d)] via a transition state which again extensively delocalizes the

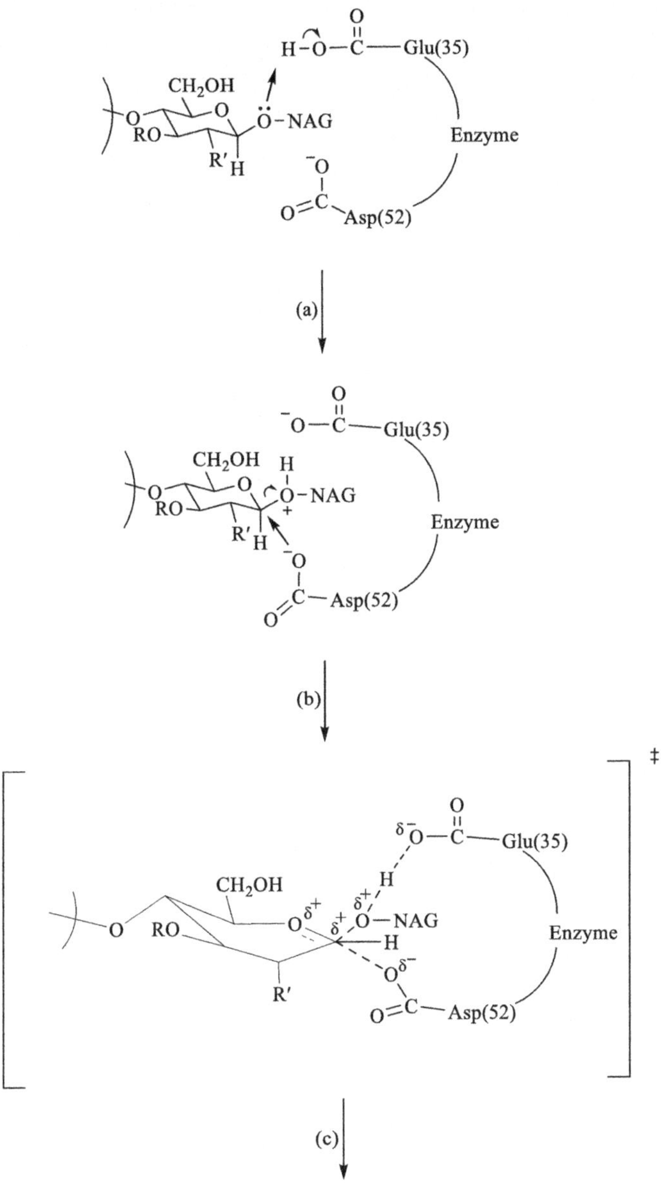

FIGURE 22.6 Detailed mechanism for lysozyme-catalyzed hydrolysis showing the involvement of key amino acids at the active site.

developing positive and negative charges [step (e)]. The collapse of this transition state [step (f)] regenerates the active enzyme and forms the hemiacetal, again with inversion. The result is a hydrolytic reaction that has resulted in overall net retention.

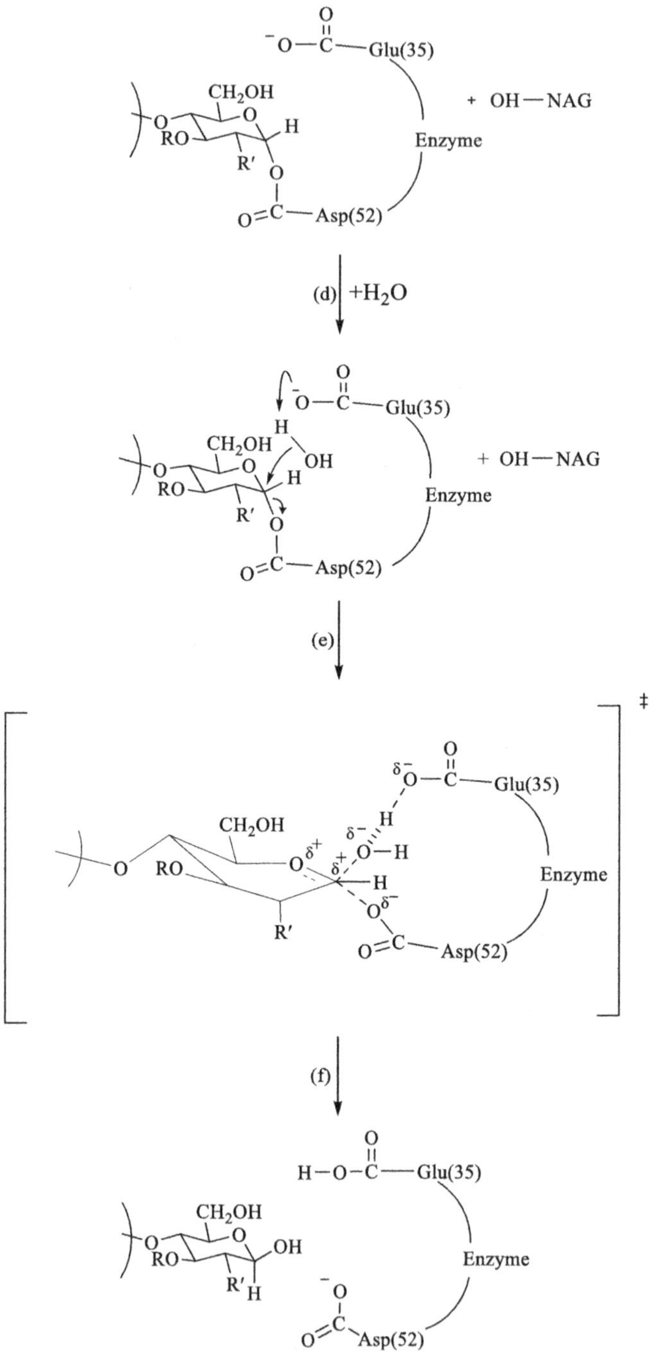

FIGURE 22.6 (Contineued).

Leading references

Coines, J.; Raich, L.; Rovira, C. *Curr. Opin. Chem. Biol.* **53**, 183–191 (2019); Ardèvol, A. and Rovira, C. *J. Am. Chem. Soc.* **137**, 7528–7547 (2015); Lombard, V.; Ramulu, H. G.; Drula, E.; Coutinho, P. M.; Henrissat, B. *Nuclei Acids Res.* **42**, D490–D495 (2014); Mhlongo, N. N.; Skelton, A. A.; Kruger, G.; Soliman, M. E. S.; Williams, I. H. *Proteins* **82**, 1747–1755 (2014); Davies, G. J.; Planas, A.; Rovira, C. *Acc. Chem. Res.* **45**, 308–316 (2012); Williams, I. H.; Pernia, J. R.; Tuñón, I. *Pure Appl. Chem.* **83**, 1507–1511 (2011); Vocadlo, D. J. and Davies, G. J. *Curr. Opin. Chem. Biol.* **12**, 539–555 (2008); Strynadka, N. C. J. and James, M. N. G. *J. Mol. Biol.* **220**, 401–424 (1991); Canfield, R. E. and Liu, A. K. *J. Biol. Chem.* **240**, 1997–2002 (1965).

Chapter 23

Methyl-coenzyme M reductase

23.1 Methyl-coenzyme M reductase

Methyl-coenzyme M reductase (EC 2.8.4.1; MCR, coenzyme B sulfoethylthiotransferase) is the final enzyme involved in methanogenesis (methane biogenesis). The starting materials in this sequence are typically single-carbon species (e.g., CO_2, CH_3OH, and HCO_2^-), though acetate is also a substrate. The organisms responsible for this chemistry belong to the domain "*Archaea*" and are referred to as "methanoarchaea." The substrates for MCR are methyl-coenzyme M (2-(methylthio)ethanesulfonate; CoMS-CH_3) and coenzyme B (*N*-7-mercaptoheptanoylthreonine phosphate; CoB_7SH; CoBSH). The nickel porphinoid, coenzyme F_{430} (F_{430}), is a requisite cofactor. The structures for these compounds are shown in Fig. 23.1.

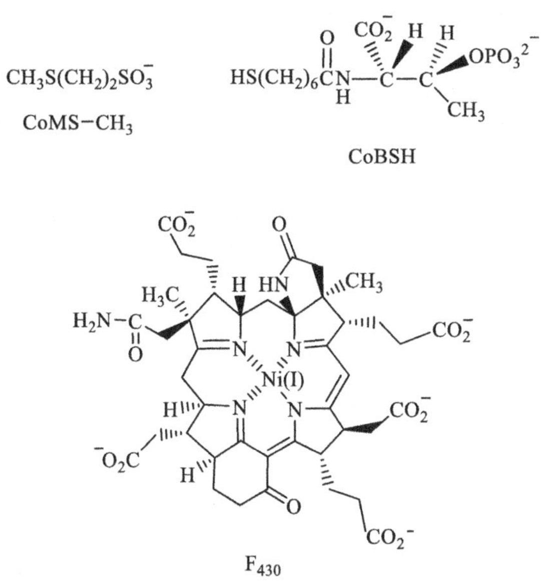

FIGURE 23.1 Structures of CoMS-CH_3, CoBSH, and F_{430}.

Enzyme Active Sites and their Reaction Mechanisms. DOI: https://doi.org/10.1016/B978-0-12-821067-3.00023-4

The chemistry catalyzed by MCR is shown in Fig. 23.2. The products are methane and a heterodisulfide (CoM-S-S-CoB). The disulfide is ultimately enzymatically reduced back to the coenzymes by hydrogen via a heterodisulfide reductase/hydrogenase system. MCR also catalyzes the oxidation of methane to form CoMS-CH_3.

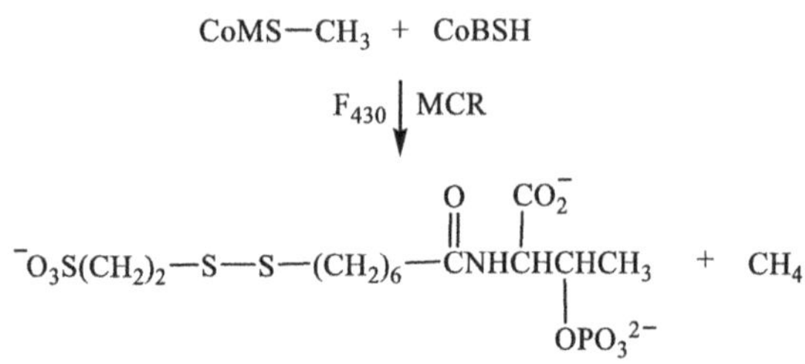

FIGURE 23.2 The overall chemistry catalyzed by MCR. *MCR*, Methyl-coenzyme M reductase.

23.2 Physiological function

Methanogenesis is an anaerobic, enzymatic sequence that is responsible for all biologically generated methane on earth.

23.3 Key structural features

MCR is a multiprotein complex consisting of two identical halves, each with an active site. The two sites are approximately 50 Å apart and contain the F_{430} cofactor. The active form of the enzyme utilizes the F_{430} nickel in its Ni(I) oxidation state (often referred to as the MCR_{red1} state). The active site lies at the bottom of a funnel-shaped channel about 50 Å in length, with an approximately 25 Å diameter top and an approximately 8 Å diameter bottom. The F_{430} lies at the bottom of the channel. A glutamine residue (Gln_{147} in *Methanobacterium thermoautotrophicum*) binds to the nickel via its side chain amide oxygen on the rear face, and the sulfur atom of CoMS-CH_3 binds to the front (active) face. The substrates bind sequentially—CoMS-CH_3 first and then CoBSH. Substrate binding induces conformational changes that bring the three reactants closer to one another and seals the channel from the aqueous environment.

The channel provides a hydrophobic environment, with several highly conserved residues (Phe_{330}, Phe_{361}, and Phe_{443}) that envelop the SH tail of CoB, and forms H-bonds (Tyr_{333} and Tyr_{367}) to the sulfur atom of CoMS-CH_3. The sulfonate of CoMS-CH_3 forms a salt bridge to the guanidinium ion of Arg_{120} and H-bonds to the peptide nitrogen of Tyr_{444} and to a water molecule (the latter also forms an H-bond with the peptide oxygen of His_{364}). The portion of the active site in the vicinity of Fe_{430} is shown in Fig. 23.3.

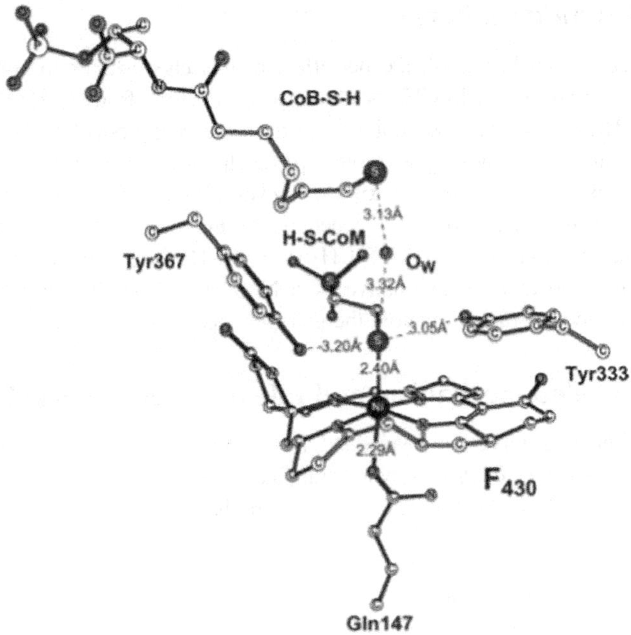

FIGURE 23.3 The Fe_{430} region of the active site of MCR from *M. thermoautotrophicum.* *Reprinted with permission from Springer: Pelmenschikov, V.; Siegbahn, P. E. M. J. Biol. Inorg. Chem. 8, 653–662 (2003); Copyright 2003.*

A novel feature of MCR is the presence of four modified amino acid residues containing methyl groups, two of which (2-(S)-methylglutamine and 5-(S)-methylarginine) having been previously unobserved (see Fig. 23.4). A fifth novel AA, thioglycine, is also present in the protein. It was shown that the methyl groups derive from *S*-adenosyl-L-methionine (see Chapter 12: m^5C Cytosine Methyltransferase).

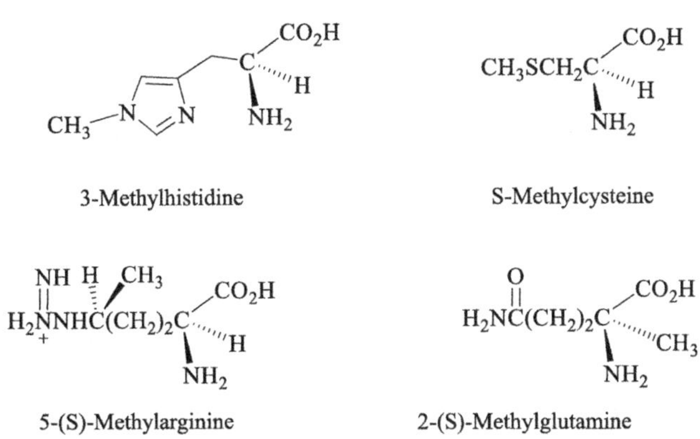

FIGURE 23.4 Modified amino acid residues found in MCR. *MCR*, Methyl-coenzyme M reductase.

23.4 Reaction sequence

As is evident from Fig. 23.2, the net effect of this chemistry is to combine the methyl group from CoMS-CH_3 with a hydrogen atom from CoBSH to form methane. The thiyl residues combine to form the second product, CoM-S-S-CoB. The key issues in developing a (multistep) mechanism for these transformations have been determining the role of the nickel catalyst and more specifically, whether a methyl-nickel bond is generated during the reaction. Computations indicate that the conversion of an S-CH_3 to a Ni-CH_3 species would be energetically unfavorable and the alternative of a Ni-S bond, and the formation of an intermediate methyl radical, is now the generally accepted pathway.

23.5 Detailed mechanism and role of active site residues

Perhaps of greatest impact are Tyr_{333} and Tyr_{367}. Their hydrogen bonds to the CoM sulfur atom facilitate the reaction by lowering the energy of sequential transition states involving a negative charge on this sulfur. The mechanism is outlined in Fig. 23.5. In

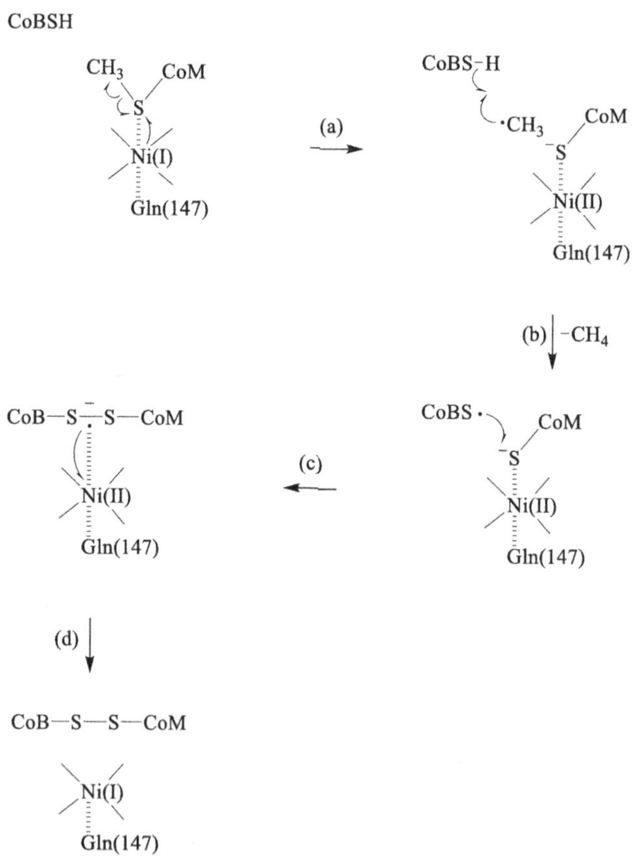

FIGURE 23.5 Detailed mechanism for MCR. *MCR*, Methyl-coenzyme M reductase.

step (a) homolytic cleavage of the CoMS-methyl bond, concomitant with oxidation of the Ni(I) species to Ni(II) [often referred to as the $MCR_{ox1\text{-}silent}$ (i.e., no EPR signal) state], leads to a methyl radical and the thiolate anion. The methyl radical abstracts a hydrogen atom from CoBSH in step (b), thus forming methane and the thiyl radical, $CoBS^{\bullet}$. The thiyl radical bonds with the CoM thiolate anion in step (c) to form the heterodisulfide radical anion, which in step (d) transfers an electron to the nickel to regenerate Ni(I) and form the neutral disulfide product.

Leading references

Thauer, R. K. *Biochemistry* **58**, 5198–5220 (2019); Siegbahn, P. E. M.; Chen, S.-L.; Liao, R.-Z. *Inorganics* **7**, 95–124 (2019); Ragsdale, S. W.; Raugei, S.; Ginovska, B.; Wongnate, W. *RSC Metallobiol. Ser.* **10**, 149–169 (2017); Wongnate, T.; Sliwa, D.; Ginovska, B.; Smith, D.; Wolf, M. W.; Lehnert, N.; Raugel, S.; Ragsdale, S. W. *Science* **352**, 953–958 (2016); Wongnate, T. and Ragsdale, S. W. *J. Biol. Chem.* **290**, 9322–9334 (2015); Blomberg, M. R. A.; Borowski, T.; Himo, F.; Liao, R.-Z.; Siegbahn, P. E. M. *Chem. Rev.* **114**, 3601–3658 (2014); Pelmenschikov, V. and Siegbahn, P. E. M. *J. Biol. Inorg. Chem.* **8**, 653–662 (2003); Thauer, R. K. *Microbiology* **144**, 2377–2406 (1998).

Chapter 24

Methylmalonyl coenzyme A mutase

24.1 Methylmalonyl coenzyme A mutase

Methylmalonyl coenzyme A mutase (EC 5.4.99.2; MCM; MUT) is an enzyme which has, as its primary function, the (reversible) conversion of (*R*)-methylmalonyl-CoA (MM-CoA) to succinyl-CoA (see Figs. 24.1

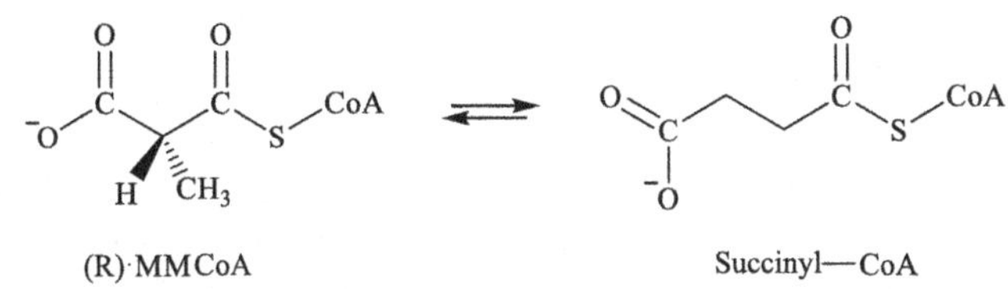

FIGURE 24.1 Chemistry catalyzed by MCM. *MCM*, Methylmalonyl coenzyme A mutase.

and 24.2). The enzyme requires vitamin B_{12} (5′-deoxyadenosylcobolamin; AdoCbl) as a cofactor (Fig. 24.3). There are three classes of reactions catalyzed by AdoCbl: (I) carbon skeleton rearrangements, (II) elimination reactions, and (III) aminomutase reactions. MCM falls into class I, along with

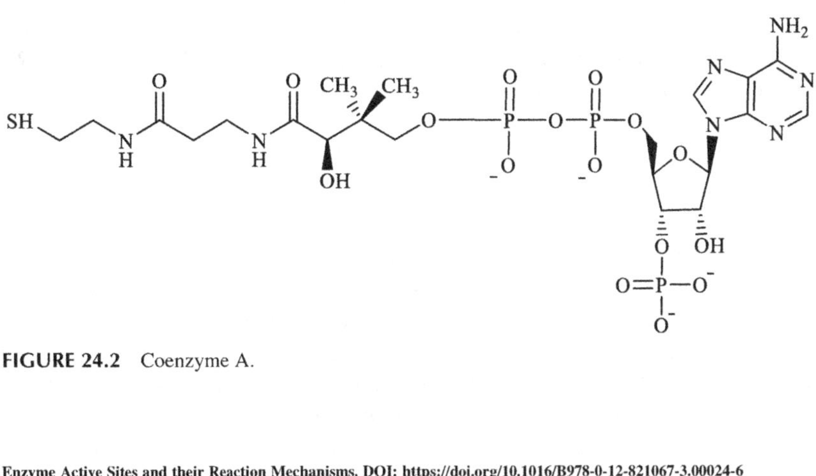

FIGURE 24.2 Coenzyme A.

Enzyme Active Sites and their Reaction Mechanisms. DOI: https://doi.org/10.1016/B978-0-12-821067-3.00024-6

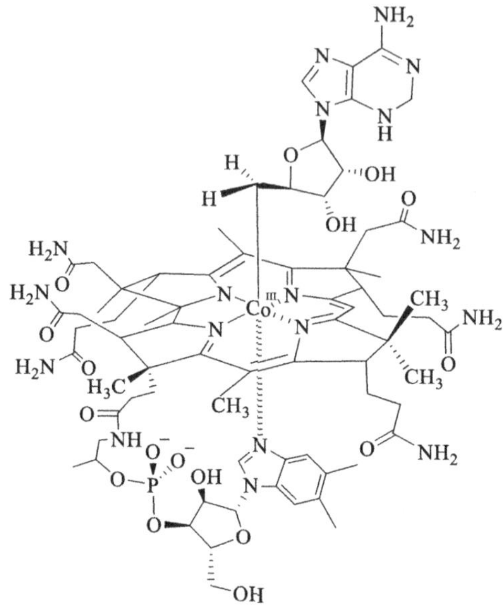

FIGURE 24.3 5′-Deoxyadenosylcobolamin.

glutamate mutase, 2-methyleneglutarate mutase, isobutyl-CoA mutase, and ethylmalonyl-CoA mutase. Of this group, only MCM is found in mammals. The only other Cbl-dependent enzyme found in humans is methionine synthase, which utilizes methylcobolamin.

24.2 Physiological function

MCM is a mitochondrial enzyme found in the kidney, liver, heart, and throughout the central nervous system. Animals cannot synthesize AdoCbl, and its absence in the diet leads to pernicious anemia and degeneration of the nervous system. MCM is an essential enzyme in the metabolic pathways for branched-chain amino acids, odd-chain fatty acids, and cholesterol. (*R*)-MM-CoA is derived from propionyl-CoA; succinyl-CoA is an intermediate within the citric acid cycle.

24.3 Key structural features

The mammalian enzyme exists as a homodimer with each monomer having two domains—a large one that binds the substrate, and a smaller one that binds the cofactor. The active site is positioned at the interface of the two domains. A metallochaperone protein (methylmalonic aciduria type A) associates with human MCM in a protective mode during MCM catalysis. There are numerous AAs involved in binding the cofactor and the substrate.

A number of these assist in the homolytic cleavage of the dAdo − Co bond, the stabilization and protection of the resultant dAdo radical and the lowering of the activation energy for the skeletal rearrangement. Though an X-ray structure for human MCM has been published, more detailed analyses exist for the enzyme from *Propionibacterium shermanii*. Thus the figures that follow are from X-ray structures of the *P. shermanii* active site, with the human equivalent AA residue numbering noted when possible.

Figs. 24.4 and 24.5 show the binding sites for desulpho-CoA and coenzyme B_{12}, respectively. Fig. 24.6 shows the binding of AdoCbl, and Fig. 24.7 shows bound substrate and its two most important interacting residues.

Noteworthy:

1. In the lower right-hand portion of Fig. 24.5, one finds a triad of His_{610}, Asp608, and Lys604 (bacterial; human equivalent- His627, Asp625, and Lys621). The histidine has broken the Co-dimethylbenzimidazole (DMB) bond prior to catalysis; the replacement of the DMB "tail" by His classifies MCM as a "base-off/His-on" enzyme. The Asp and Lys provide hydrogen bonding that stabilizes the Co-His axial linkage as it elongates during the homolytic reaction (see mechanism below).
2. Also in Fig. 24.5, we find Tyr_{89} (bacterial) lies above, and stacks with, the dAdo group. It is substantially displaced upon the binding of substrate into the active site. This is thought to play an important role in facilitating the Ado-Co homolysis, in reorienting the dAdo radical during the rearrangement and in steering the stereochemistry of the hydrogen abstraction reaction.

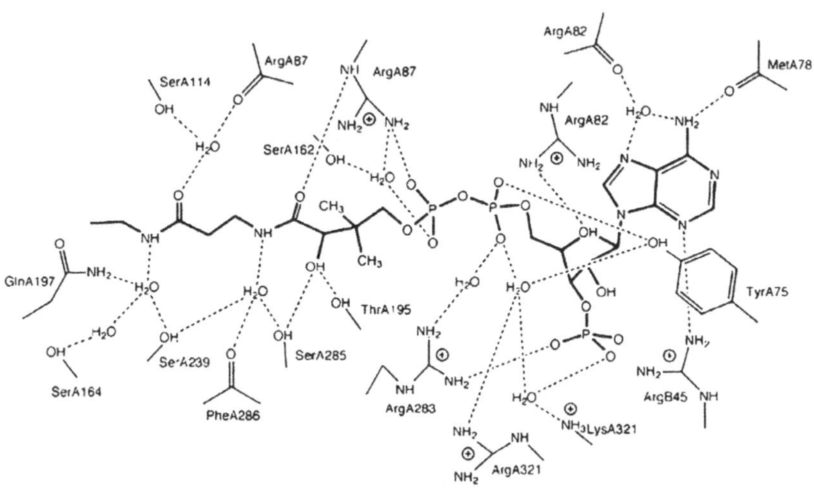

FIGURE 24.4 The AA interactions of a CoA analog lacking the terminal sulfur contained within the *Propionibacterium shermanii* active site. *AA*, amino acid; *CoA*, coenzyme A. *From Mancia, F. et al. How coenzyme B_{12} radicals are generated: the crytal structure of methylmalonyl coenzyme A mutase at 2 Å resolution,* Structure, *4, 339–350 (1996).*

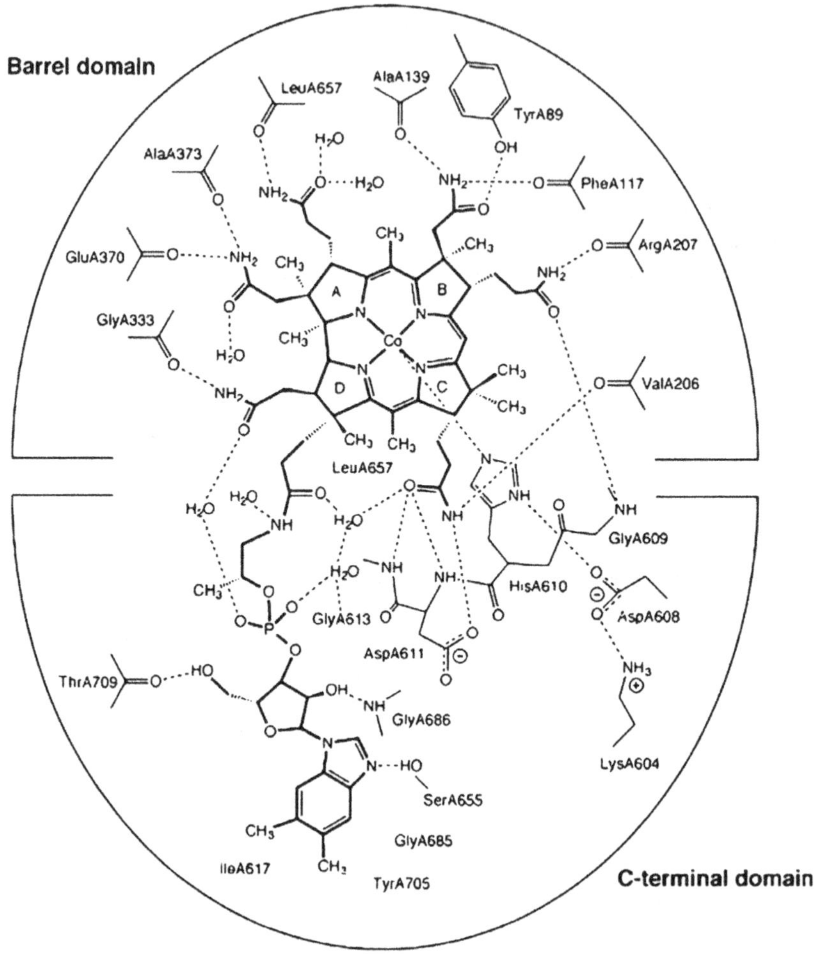

FIGURE 24.5 The AA interactions of coenzyme B_{12} contained within the *Propionibacterium shermanii* active site. *AA*, amino acid. *From: Mancia, F. et al. (1996). See Fig. 24.4.*

3. In Fig. 24.6, we find that the Glu_{370} (bacterial; human equivalent-Glu_{392}) has formed hydrogen bonds with the 2′ and 3′-hydroxyl groups of the AdoCbl ribose ring. These are essential for both the initial hemolysis and a consequential reorientation of the Ado radical to place it in proximity to the substrate hydrogen.
4. As regard the substrate, in Fig. 24.7, His_{244} (bacterial) hydrogen bonds to the carbonyl group of the thioester, thus stabilizing the intermediate cyclopropyloxy radical during the rearrangement. There is also evidence that this histidine shields the dAdo radical from molecular oxygen. Arg_{207} (bacterial) hydrogen bonds to the substrate carboxylate group.

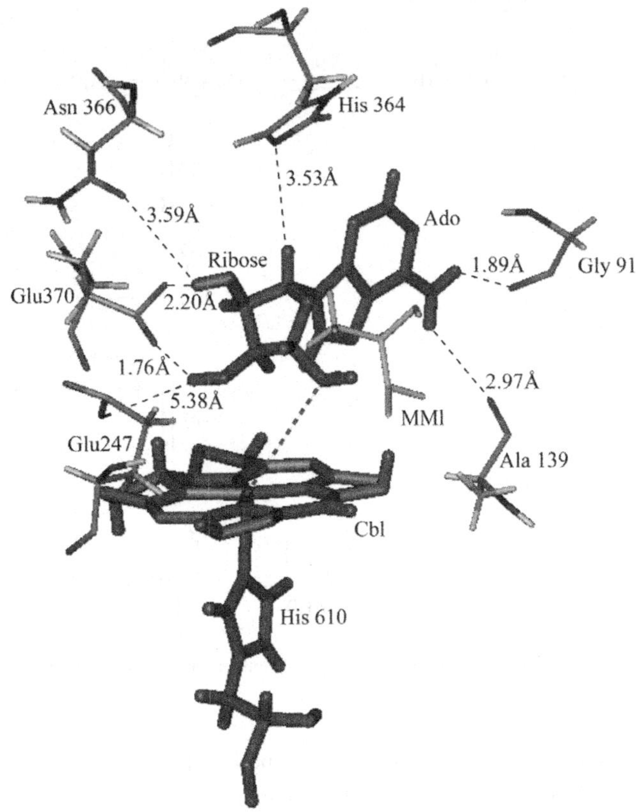

FIGURE 24.6 A close-up view of key residues with AdoCbl bound into the active site. *From Sharma, P. K. et al.* Proc. Natl. Acad. Sci. *104, 0661–9666 (2007). Copyright 2007 National Academy of Sciences, USA.*

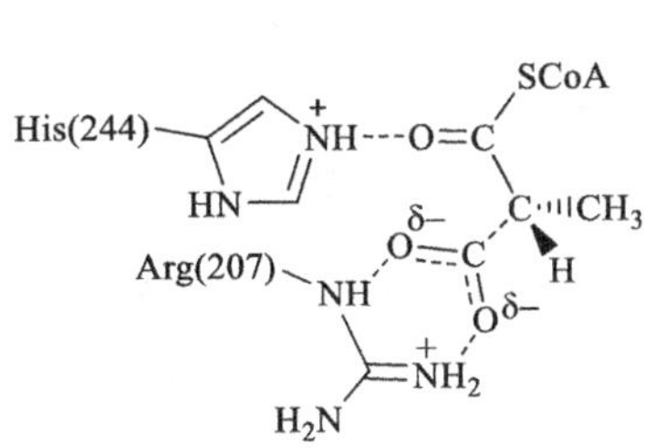

FIGURE 24.7 Substrate interactions in the active site with His_{244} and Arg_{207}. *Adapted with permission from Banerjee, R.* Chem. Rev. *103, 2083–2094 (2003). Copyright 2003, American Chemical Society.*

24.4 Reaction sequence

The skeletal rearrangement catalyzed by MCM proceeds by a free radical mechanism, a relatively uncommon process in organic

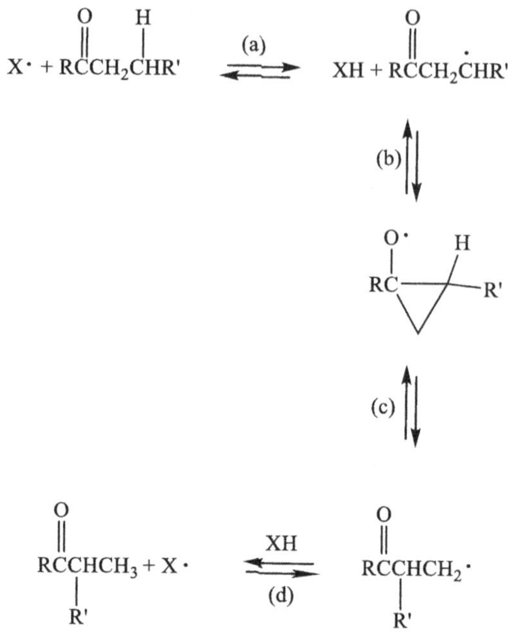

FIGURE 24.8 Chemical reaction sequence for the conversion of (*R*)-methylmalonyl-CoA to succinyl-CoA.

chemistry. The general sequence is outlined in Fig. 24.8. In step (a), an initiating radical abstracts a hydrogen atom from the reactant, selectively β to the carbonyl group, a level of specificity unique to the enzyme. Note the use of single-headed arrows to denote one-electron reactions. Step (b) involves attack by the newly generated alkyl radical on the pi system of the carbonyl group to form a cyclopropyloxy radical. The C − C bond originally α to the carbonyl group is then homolytically cleaved in step (c) to reopen the three-membered ring, and the resulting alkyl radical abstracts a hydrogen atom from XH to complete the reaction. Note that all these steps are reversible as, in fact, is the enzymatic chemistry (i.e., succinyl-CoA can be converted by MCM into methylmalonyl-CoA).

24.5 Detailed mechanism and the role of active site residues

The detailed sequence is presented in Fig. 24.9 with bacterial MCM depicted in cartoon fashion. Glu_{370} is included because it is believed to be critical, not only in helping to stabilize the 5′-dAdo radical but also in steering this species toward the MM-CoA methyl group for the next step. Though not explicitly depicted here, significant conformational changes in the cofactor occur when the substrate joins it in the active site. Cleavage of the Ado-Co bond in

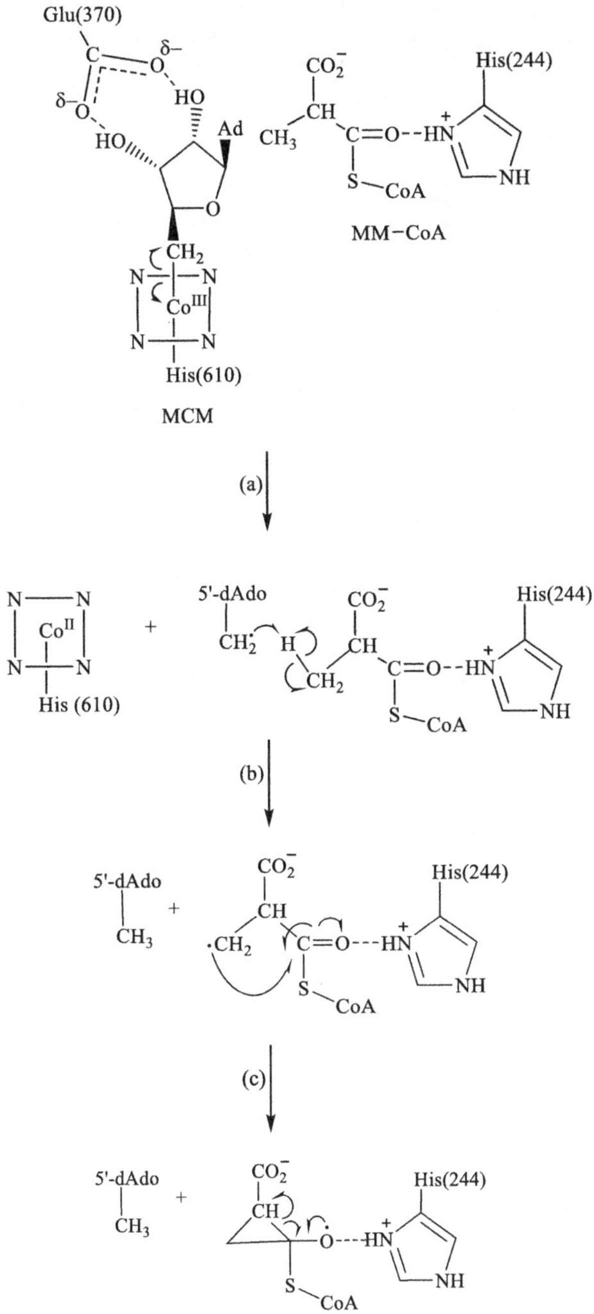

FIGURE 24.9 Detailed mechanism for methylmalonyl coenzyme A mutase showing the involvement of key amino acids at the active site.

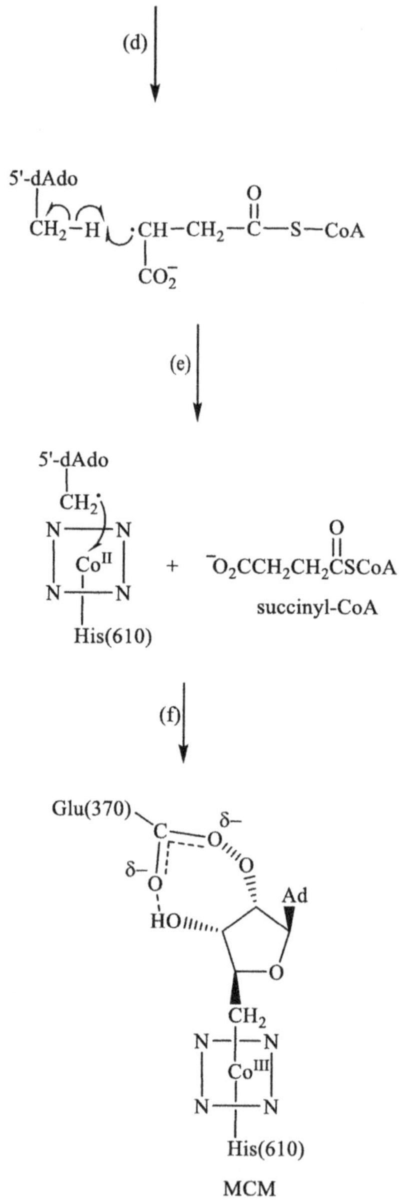

FIGURE 24.9 (Continued).

step (a) occurs with a rate enhancement of 10^{12} when the substrate is present and with the "base-off/His-on" structure. Note the one-electron reduction of Co^{III} to Co^{II} and the use of "single-headed arrows" in this cleavage.

The 5′-dAdo radical abstracts a hydrogen atom from the MM-CoA methyl group in step (b). Though the distance between these species precludes a concerted cleavage/abstraction process, isotope effect studies and computations indicate that "hydrogen tunneling" is involved. In a "tunneling" mechanism, the migrating (H) atom moves "through", rather than over, an activation energy barrier. Step (c) involves addition to the carbonyl group, aided by the hydrogen bond to His_{244}. In step (c), the three-membered ring homolytically cleaves at the C−C bond originally alpha to the thioester carbonyl carbon. The new substrate radical abstracts a hydrogen atom from the 5′-dAdo methyl group in step (d), and the cofactor reforms in step (e).

Leading references

Sokolovskaya, O. M.; Mok, K. C.; Park, J. D.; Tran, J. L. A.; Quanstrom, K. A.; Taga, M. E. *mBio* **10**, 1–17 (2019); Conrad, K. S.; Jordan, C. D.; Brown, K. L.; Brunhold, T. C. *Inorg. Chem.* **54**, 3736–3747 (2015); Makins, C.; Pickering, A. V.; Mariani, C.; Wolthers, K. R. *Biochemistry* **52**, 878–888 (2013); Jones, A. R.; Levy, C.; Hay, S.; Scrutton, N. S. *FEBS J.* **280**, 2997–3008 (2013); Dowling, D. P.; Croft, A. K.; Drennan, C. *Annu. Rev. Biophys.* **41**, 403–427 (2012); Marsh, E. N. G. and Meléndez, G. D. R. *Biochim. Biophys. Acta* **1824**, 1154–1164 (2012); Gruber, K.; Puffer, B.; Kräutler, B. *Chem. Soc. Rev.* **40**, 4346–4363 (2011); Sharma, P. K.; Chu, Z. T.; Olsson, M. H. M.; Warshel, A. *Proc. Natl. Acad. Sci.* **104**, 9661–9666 (2007); Banerjee, R. *Chem. Rev.* **103**, 2083–2094 (2003); Palmenschikov, V. and Siegbahn, P. E. M. J. *Biol. Inorg. Chem.* **8**, 652–662 (2003); Mancia, F.; Keep, N. H.; Nakagawa, A.; Leadley, P. F.; McSweeney, S.; Rasmussen, B.; Bösecke, P.; Diat, O.; Evans, P. R. *Structure* **4**, 339–350 (1996).

Chapter 25

Nonheme iron halogenase

25.1 Syringomycin halogenase

Syringomycin halogenase (SyrB2) is one of a family of mononuclear nonheme iron (NHFe) enzymes that is α-ketoglutarate (αKG), chloride, and dioxygen dependent. It is also one member of a larger family of halogenases that employ multiple mechanisms and cofactors to halogenate substrates (see also Chapter 40: Vanadium-dependent Chloroperoxidase). In the case of SyrB2, the substrate is L-threonine, which is chlorinated regiospecifically on its γ-carbon to form 4-chloro-L-threonine. The net consequence of the multistep reaction is shown in Fig. 25.1.

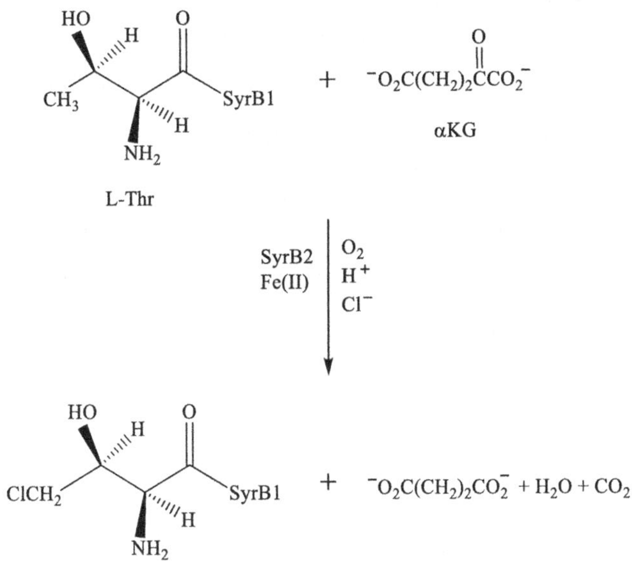

FIGURE 25.1 The overall chemistry catalyzed by SyrB2. *SyrB2*, Syringomycin halogenase.

Note that though Fig. 25.1 is seemingly complex, the equation is balanced by charge and atoms. Since the conversion of a C—H bond to a C—Cl bond is effectively an "oxidation," there must be a concomitant reduction (here of O_2). The iron is catalytic; it begins and ends the chemistry as Fe^{2+}. The threonine is actually "delivered" to the active site by an "acyl carrier protein," SyrB1, in

Enzyme Active Sites and their Reaction Mechanisms. DOI: https://doi.org/10.1016/B978-0-12-821067-3.00025-8

the form of a thioester of phosphopantatheine (see Fig. 25.2). It remains tethered in this way throughout the halogenation process.

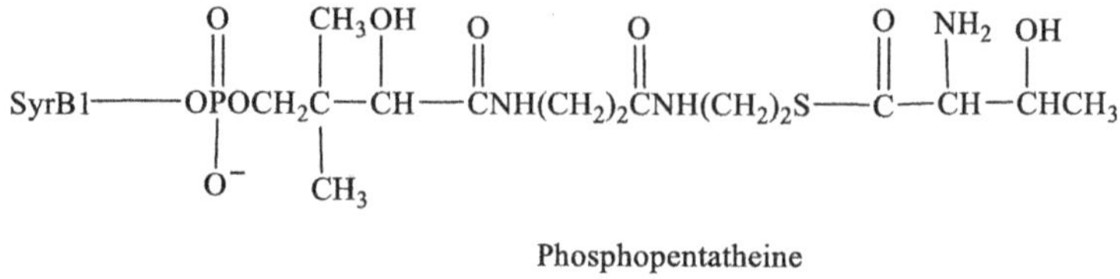

FIGURE 25.2 Phosphopantatheine.

25.2 Physiological function

SyrB2 is a bacterial enzyme isolated from *Pseudomonas syringae*. Chlorinated L-threonine is an intermediate in the biosynthesis of a phytotoxin, syringomycin E.

25.3 Key structural features

As will be seen below, the chemistry of SyrB2 resembles that of a large class of nonheme iron enzymes that can replace a nonactivated C—H bond with a hydroxyl group. These hydroxylases also require αKG and utilize molecular oxygen. The central atom in the hydroxylase active site is Fe(II), to which is bound αKG, a water molecule, and an AA triad consisting of two histidine and one aspartate (or glutamate) residues. The overall geometry is that of an octahedral complex. The bonding to the iron atom in the SyrB2 active site (Fig. 25.3) differs only in the replacement of Asp by a chlorine atom. Note the stabilization of the αKG carboxylate by Thr_{113}, Arg_{248}, Trp_{145}, and Phe_{104}. The chlorine atom lies in a hydrophobic pocket made up of Ala_{118}, Phe_{121}, and Ser_{231}. The Ala_{118}, in effect, replaces the more typical Asp of the hydrolase triad and provides the space needed for the chloride binding to Fe. Arg_{254} interacts with the apical water ligand. Fig. 25.3 omits, for the sake of clarity, two water molecules that lie near, and interact with, the chlorine atom. One of the water molecules also interacts with Thr_{143} and the other with Asn_{123}.

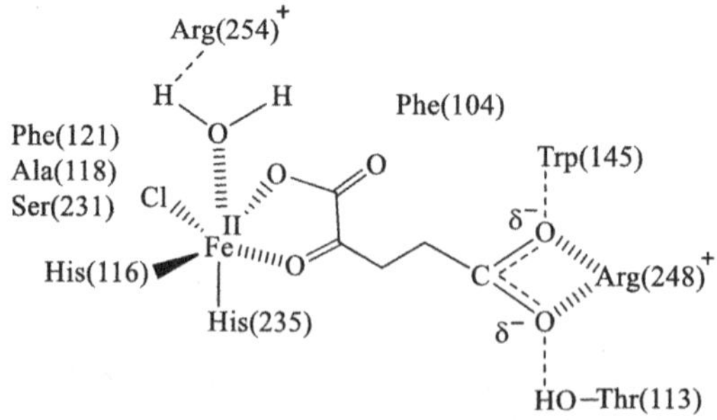

FIGURE 25.3 The SyrB2 active site. *SyrB2*, Syringomycin halogenase. *Adapted by permission from Springer: Nature, Crystal structure of the non-haem iron halogenase SyrB2 in syringomycin biosynthesis, Blasiak, L. C. et al., Copyright 2006.*

25.4 Reaction sequence

The elucidation of the mechanism of SyrB2 has benefited from two factors: (1) the comprehensive body of knowledge that has accumulated on the mechanistic details for the analogous hydroxylase family and (2) a very extensive body of theoretical analyses of the reaction sequence. The generally accepted sequence is shown in Fig. 25.4. The chemistry starts with the entry of the SyrB1-tethered amino acid ($R{-}CH_3$) concomitant with the release of the apical water. This sequence of substrate binding and ligand release has been referred to as "substrate triggering." O_2 enters the active site and the subsequent release of carbon dioxide generates a ferryl (Fe=O) species (note the new stereoelectronic configuration about the central iron). This is followed by H-abstraction from the Thr-CH_3 group and the transfer of the iron-bound chlorine atom to the methylene radical. Succinate is released and ultimately the addition of a chloride ion and the binding of αKG reforms the active enzyme.

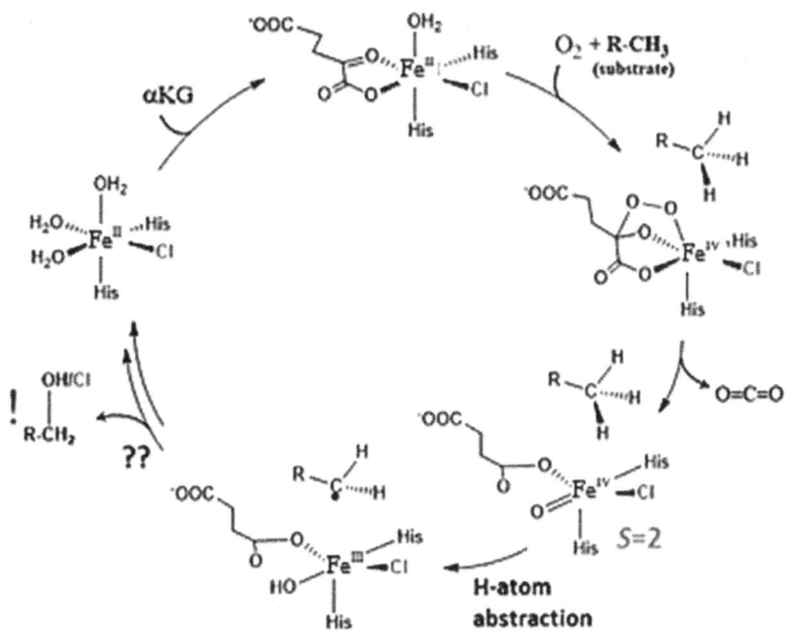

FIGURE 25.4 Reaction sequence for the chlorination of L-Thr by SyrB2. *SyrB2*, Syringomycin halogenase. *Reprinted with permission from Srnec, M. and Solomon, E. J.* J. Am. Chem. Soc. *139, 2396–2407 (2017). Copyright 2017, American Chemical Society.*

25.5 Detailed mechanism and the role of active site residues

A more detailed mechanism is presented in Fig. 25.5. This figure has been constructed primarily to help the reader track the bond-making and bond-breaking steps by the use of arrows *while maintaining the conservation of charge throughout*. Resonance contributors (double-headed arrows) for several of the intermediates are used to facilitate tracking the net chemical transformations.

The mechanism starts with step (a) after the substrate (now abbreviated to Thr-L-CH_3) has triggered the loss of the apical water. This step is facilitated by H-bonding of the Thr-OH group to Glu_{102}. The binding of molecular oxygen into the vacated site leads to an intermediate for which the oxidation state for the Fe can be drawn as Fe(II), Fe(III), or Fe(IV). Corresponding changes in the number of electrons on the oxygen are required to keep the overall charge of the two species at +2. Cyclization (step (b)) and the expulsion of carbon dioxide (step (c)) generate the mono-oxygenated, ferryl species, which can again be represented by structures with multiple iron and oxygen oxidation states (see also Chapter 20: Indoleamine 2,3-Dioxygenase-1 and Chapter 11: Cytochrome P450cam). A hydrogen atom abstraction (step (d)), and the subsequent transfer of the chlorine atom from the iron to the alkyl radical (step (e)) generates 4-chloro-L-threonine. Loss of the succinate group (step (f)) and the subsequent addition of αKG regenerate the catalytic complex (see Fig. 25.4).

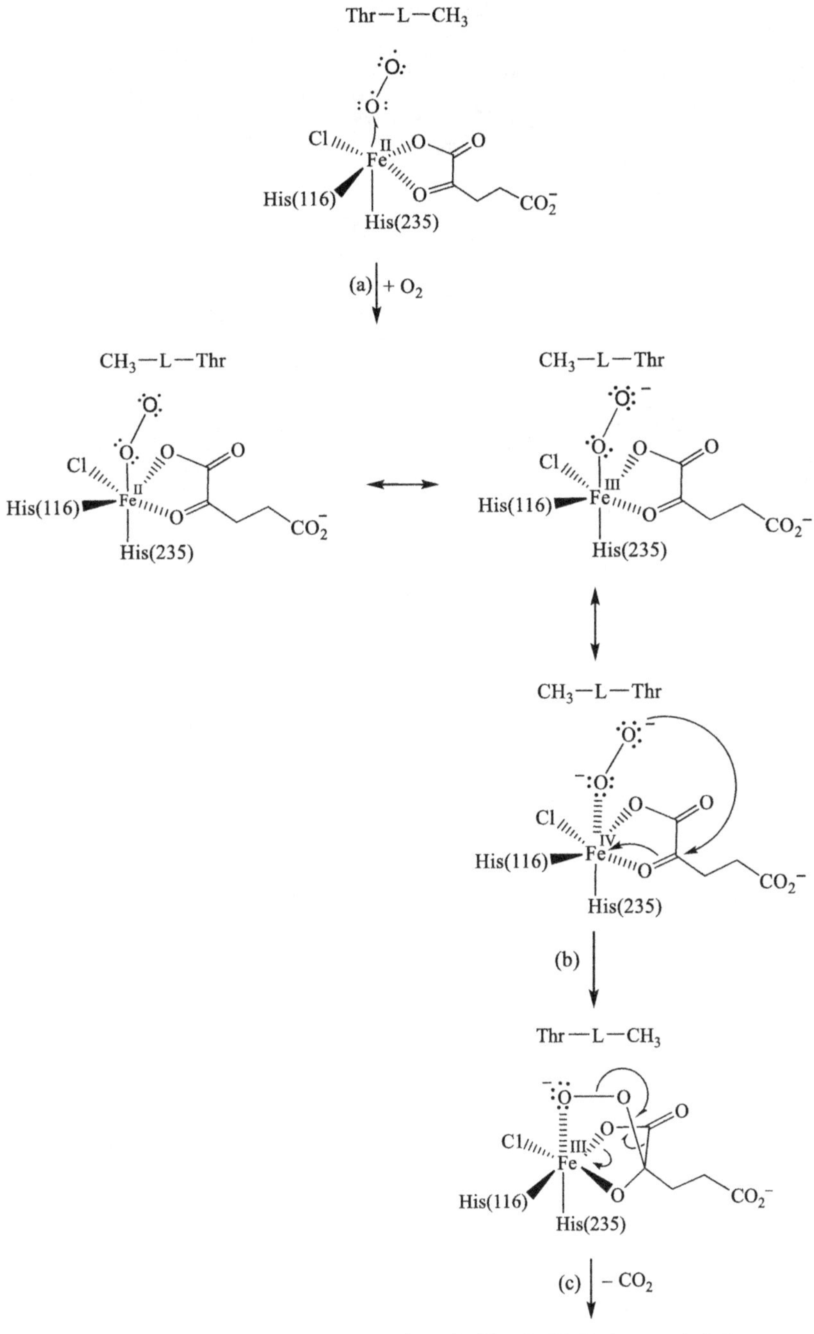

FIGURE 25.5 Mechanism for SyrB2 chlorination of L-Thr. *SyrB2*, Syringomycin halogenase.

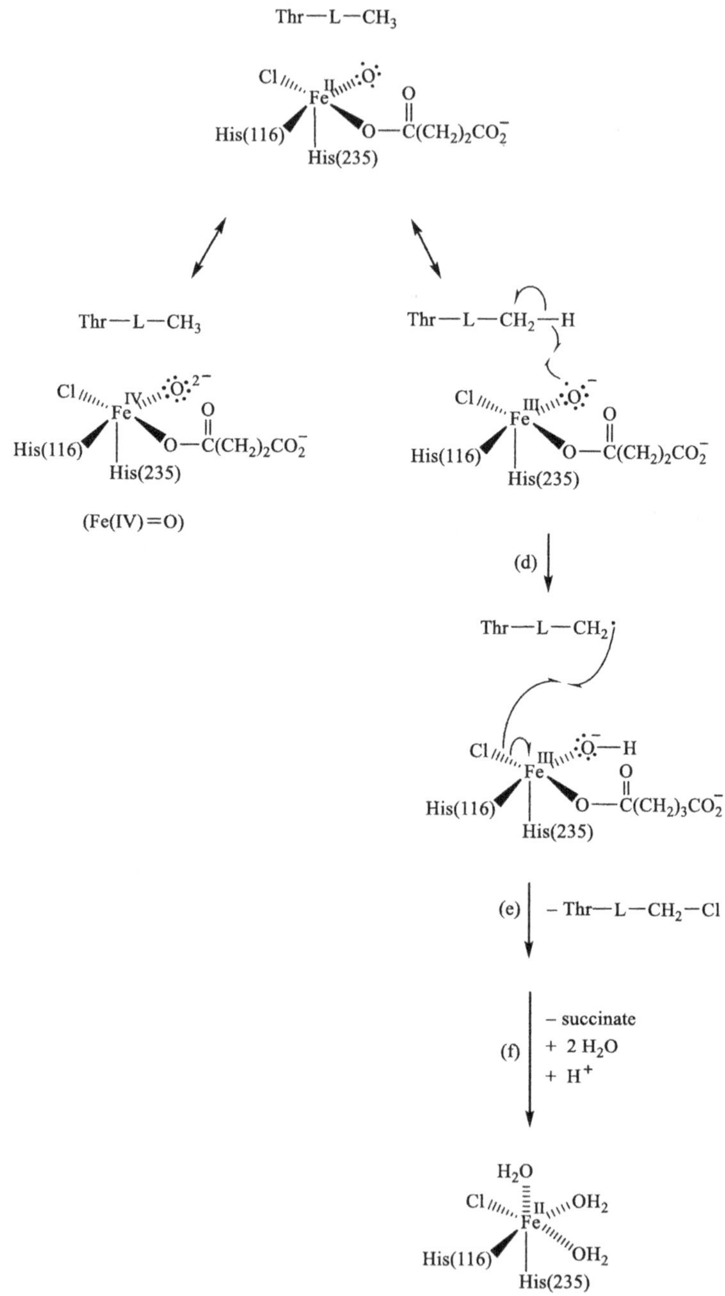

FIGURE 25.5 (Continued).

Leading references

Mehmood, R.; Qi, H. W.; Steeves, A. H.; Kulik, H. J. *Catalysis* **9**, 4930–4943 (2019); Timmins, A. and deVisser, S. P. *Catalysts* **8**, 314–338 (2018); Rugg, C. and Senn, H. M. *Phys. Chem. Chem. Phys.* **19**, 30107–30119 (2017); Srnec, M. and Solomon, E. I. *J. Am. Chem. Soc.* **139**, 2396–2407 (2017); Huang, J.; Li, C.; Wang, B.; Sharon, D. A.; Wu, W.; Shaik, S. *ACS Catal.* **6**, 2694–2704 (2016); Srnec, M.; Wong, S. D.; Matthews, M. L.; Krebs, C.; Bollinger, J. M., Jr.; Solomon, E. I. *J. Am. Chem. Soc.* **138**, 5110–5122 (2016); Blasiak, L. C. and Drennan, C. L. *Acc. Chem. Res.* **42**, 147–155 (2009); Blasiak, L. C.; Vaillancourt, F. H.; Walsh, C. T.; Drennan, C. L. *Nature* **440**, 368–371 (2006).

Chapter 26

Peptidyl arginine deiminase 4

26.1 Peptidyl arginine deiminase 4

Peptidyl arginine deiminase 4 (EC 3.5.3.15; protein arginine deiminase 4; PAD4) is a member of the pentein superfamily—a group of cysteine hydrolases that require calcium activation and hydrolyze C–N bonds. Other enzymes in this family are amidinotransferase, arginine deiminase, and dimethylarginine dimethylaminohydrolase. There are five PADs (PAD1, PAD2, PAD3, PAD4, and PAD6) that are characterized by their catalytic hydrolysis of only peptidyl arginine residues. The five PADs in humans share c.50% sequence similarity and PAD4 is found widely in mammals. The general chemical transformation is shown in Fig. 26.1. This conversion is an example of a "posttranslational modification" of a protein, a phenomenon that generates AA residues with atypical functional groups (mainly by modifying lysine, serine, and arginine AA residues).

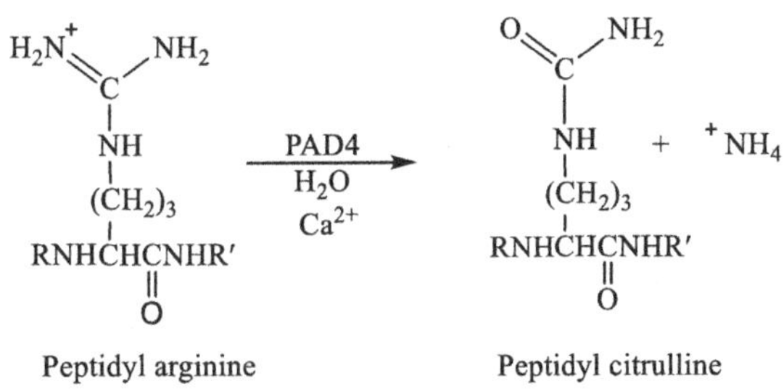

FIGURE 26.1 Formation of peptidyl citrulline from peptidyl arginine.

26.2 Physiological function

Citrullination is involved in the development and regulatory processes. PADs and citrullinated proteins are also associated with several diseases, including rheumatoid arthritis, psoriasis, and multiple sclerosis, and PAD4 is overexpressed in several cancers. PAD4 is involved in the posttranslational modification of histones by deiminating arginine in histones H2A, H3, and H4. The target arginine residues are known; for example, a single arginine within the DNA-binding site of H1 is citrullinated by PAD4. PAD4 is localized in the nucleus and is primarily found in blood neutrophils, granulocytes, and macrophages.

Enzyme Active Sites and their Reaction Mechanisms. DOI: https://doi.org/10.1016/B978-0-12-821067-3.00026-X

26.3 Key structural features

PAD4 is folded into N-terminal and C-terminal domains, with the active site located in the C-domain. It is a homodimer in its active form. A total of five calcium ions are required for activity. Interestingly, none are actively involved in the reaction mechanism. Three are distal from the active site, lie within the N-terminal domain, and help organize the overall structure of the protein for activity. Two lie at the bottom of the active site and play an essential role in creating the requisite conformational organization of the active site itself. Only after calcium binding is the key cysteine within bonding distance to initiate the chemistry (see below). The active site is a u-shaped cavity with separate entry and exit points for the substrate and for the water reactant/ammonia product. The active site of PAD4 bound to benzoyl-L-arginine amide is depicted in Fig. 26.2. Several residues have been omitted for clarity. For example, Val_{469} and Trp_{347} provide hydrophobic interactions with the 3-carbon chain. Likewise, the AA residues binding the active site calcium atoms have been omitted. The residues Asp350, His471, Asp473, and Cys645 are essential for activity and well conserved in this family.

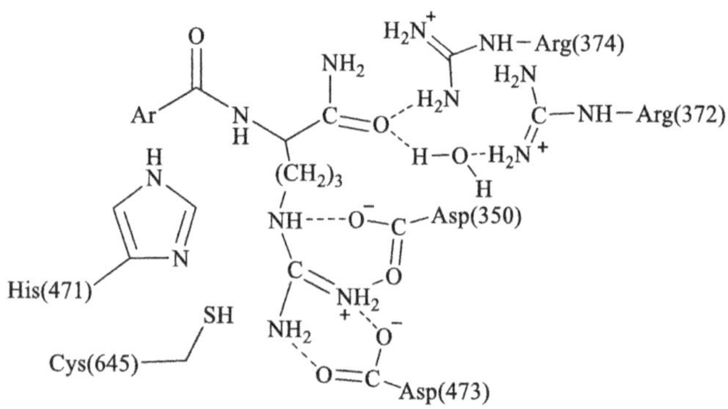

FIGURE 26.2 The active site for PAD4 with BAA as a substrate. *BAA*, Benzoyl-L-arginine amide; *PAD4*, peptidyl arginine deiminase 4. *(Adapted by permission from Springer:* Nat. Struct. Mol. Biol., *Structural basis for Ca^{2+}-induced activation of human PAD4, Arita, K. et al., Copyright 2004).*

26.4 Reaction sequence

The chemical conversion involves the hydrolysis of the "imine-like" portion of the guanidinium group to its carbonyl equivalent. The basic transformation is shown in Fig. 26.3. The entire sequence is a set of reversible reactions. The nitrogen is protonated in step (a), the nucleophilic water molecule adds to the iminium group to form the classical tetrahedral intermediate in step (b), proton transfer in step (c) activates the leaving group, and elimination in step (d) leads to the hydrolyzed product.

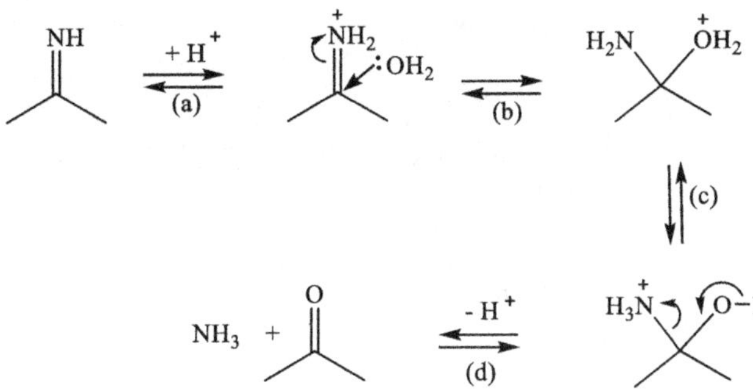

FIGURE 26.3 The addition–elimination mechanism for acid-catalyzed hydrolysis of an imine.

26.5 Detailed mechanism and the role of the active-site residues

The currently accepted mechanism for the citrullination of a peptidyl arginine (Prot-Arg in Fig. 26.4) is shown below. In step (a), the thiolate of Cys_{645} initiates the reaction through a nucleophilic addition to the guanidinium group. The distribution of charges at the outset is counterintuitive since the imidazolium group is more acidic than a thiol. A "reverse protonation" mechanism is invoked, wherein the small (c.15%) of the active site that has the requisite ionization states of Cys_{645} and His_{471} at the enzyme's optimal pH (c.7.6) is deemed responsible for the reaction to proceed. (For an excellent review of this concept, see Frankel *et al.* below). Protonation of the tetrahedral intermediate by His_{471} therefore occurs in step (b) and the expulsion of ammonia, and the entrance of water into the active site, takes place in step (c). (Note that this positions His_{471} more distal relative to Cys_{645} than is indicated in Fig. 26.2). His471-assisted deprotonation of the water facilitates nucleophilic attack by the hydroxyl on the iminium group in step (d). The reaction ends with expulsion of the thiolate and formation of the citrulline, assisted by the Asp_{473}, as shown in step (e).

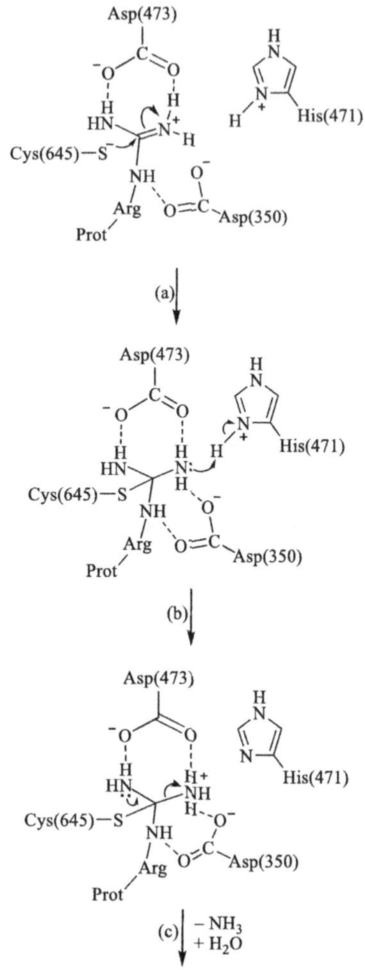

FIGURE 26.4 Detailed mechanism and the roles of active site residues. *(Adapted with permission from Fuhrmann et al.,* Chem. Rev. **115**, *5413–5461 (2015); Copyright 2015, American Chemical Society).*

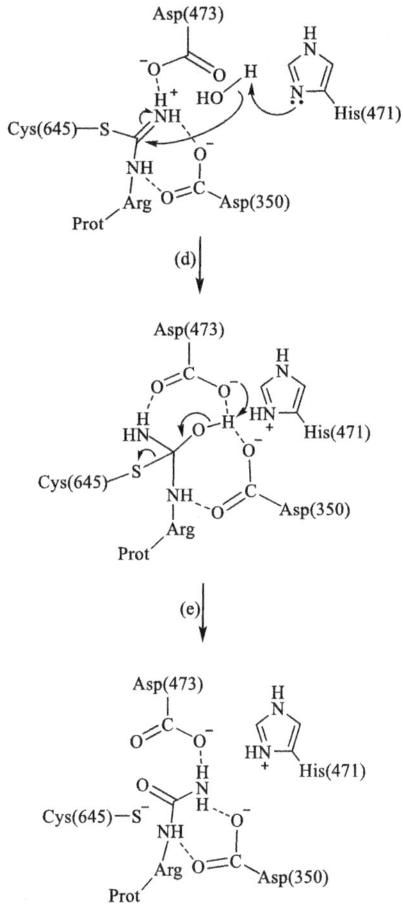

FIGURE 26.4 (Continued).

Leading references

Mondal, S. and Thompson, P. R. *Acc. Chem. Res.* **52**, 818–832 (2019);
Tjin, C. C.; Wissner, R. F.; Jamali, H.; Schepartz, A. *ACS Med. Chem. Lett.* **9**, 1013–1018 (2018); Fuhrmann, J.; Clancy, K. W.; Thompson, P. R. *Chem. Rev.* **115**, 5413–5461 (2015); Frankel, B. A.; Kruger, G. R.; Robinson, D. E.; Kelleher, N. L.; McCafferty, D. G. *Biochemistry* **44**, 11188–11200 (2005); Arita, K.; Hashimoto, H.; Shimizu, T.; Nakashima, K.; Yamada, M.; Sato, M. *Nat. Struct. Mol. Biol.* **11**, 777–783 (2004).

Chapter 27

Peptidylglycine α-hydroxylating monooxygenase

27.1 Peptidylglycine α-hydroxylating monooxygenase

Peptidylglycine α-hydroxylating monooxygenase (EC 1.14.17.3; PHM; peptidylglycine 2-hydroxylase; peptidylglycine α-hydroxylase) catalyzes hydroxylation of peptidylglycine substrates. It is one of two components of the bifunctional enzyme, peptidylglycine α-amidating monooxygenase (PAM). The second is peptidyl-α-hydroxylglycine α-amidating lyase (PAL; EC 4.3.2.5). PHM catalyzes the first step: the alpha-hydroxylation of pro-hormonal, glycine-extended peptides. This reaction is stereospecific, with the OH group replacing the *pro*-S hydrogen atom. PAL cleaves the hydroxylated product to a C-terminal carboxyamidated peptide and glyoxylate. The full sequence of the chemistry catalyzed by PAM is shown in Fig. 27.1.

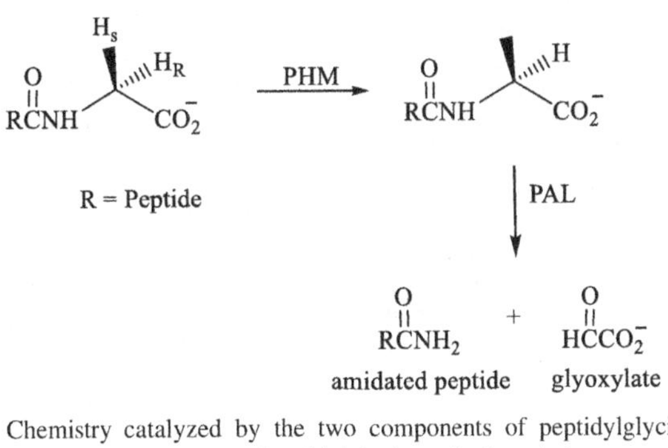

FIGURE 27.1 Chemistry catalyzed by the two components of peptidylglycine α-amidating monooxygenase—PHM and PAL. *PHM*, Peptidylglycine α-hydroxylating monooxygenase; *PAL*, peptidyl-α-hydroxylglycine α-amidating lyase.

PHM is classified as a "noncoupled binuclear copper enzyme"; related enzymes are dopamine β-monooxygenase (EC 1.14.17.1) and tyramine β-monooxygenase. The term "noncoupled" refers to the fact that two copper atoms are in the active site and essential to the chemistry. Yet, they are 11 Å apart and do not directly "communicate" with one another. "Monooxygenases" are redox enzymes that utilize one of the two O_2 oxygen atoms, with the other

Enzyme Active Sites and their Reaction Mechanisms. DOI: https://doi.org/10.1016/B978-0-12-821067-3.00027-1

converted to water. A reducing agent is required as a cofactor; ascorbate typically plays this role.

27.2 Physiological function

PAM is localized in the secretory vesicles of the pituitary gland and is required for the synthesis of bioactive peptide hormones and neurotransmitters through C-terminal amidation (e.g., oxytocin and calcitonin).

27.3 Key structural features

An X-ray structure of *N*-acetyl-diiodo-tyrosyl-D-threonine (IYT) bound into the active site of *Rattus norvegicus* is shown in Fig. 27.2. Note that the two copper atoms labeled Cu_B and Cu_A in the older literature are now labeled Cu_M and Cu_H, respectively. Tyr_{318} and Arg_{240} (not shown in Fig. 27.2) are H-bonded to the terminal carboxylate group of the substrate, and the side chain of Asn_{316} (not shown in Fig. 27.2) is H-bonded to the threonine amide N—H bond.

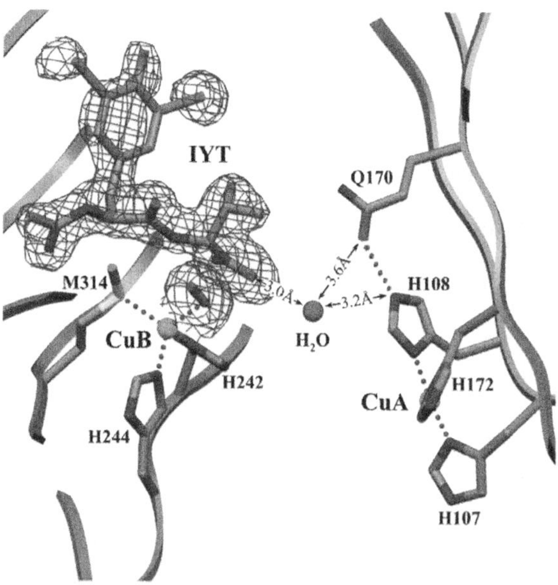

FIGURE 27.2 The active site of reduced PHM with bound *N*-acetyl-diiodo-tyrosyl-D-threonine (IYT). *PHM*, Peptidylglycine α-hydroxylating monooxygenase. *(Republished with permission of the American Association for the Advancement of Science, from Science, Prigge et al.* ***304****, 864–867 (2004); Copyright 2004 Copyright Clearance Center, Inc).*

An X-ray structure has shown that molecular oxygen becomes bound ("end on") directly to Cu_M. The copper atoms, when oxidized, are each also bound to an aquo ligand (as hydroxo for Cu_M).

The copper atoms are reduced to Cu(I) by ascorbic acid during the catalytic process. They are reoxidized to Cu(II) in the presence of the substrate and molecular oxygen. The concomitant oxidation of the ascorbate anion initially leads to the semidehydroascorbate radical and ultimately to dehydroascorbic acid (see Fig. 27.3).

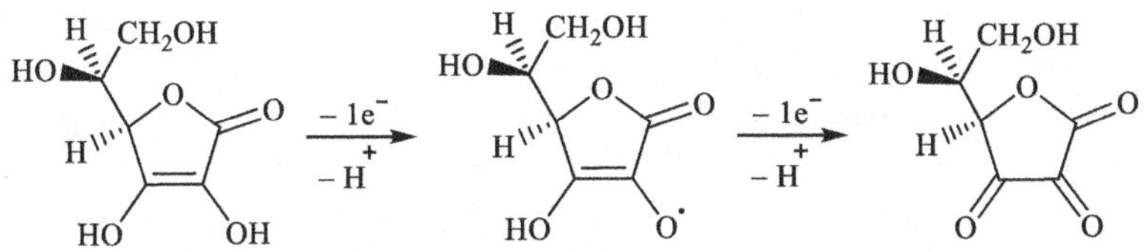

FIGURE 27.3 Stepwise oxidation of ascorbic acid to dehydroascorbic acid.

27.4 Reaction sequence

An overview of the chemistry catalyzed by PHM is shown in Fig. 27.4. For the first step, there is consensus that Cu(I) reduces dioxygen to form superoxide. The superoxide moiety then abstracts, stereospecifically, the α-H_s atom of the terminal glycine unit to form an alkyl radical and a copper-

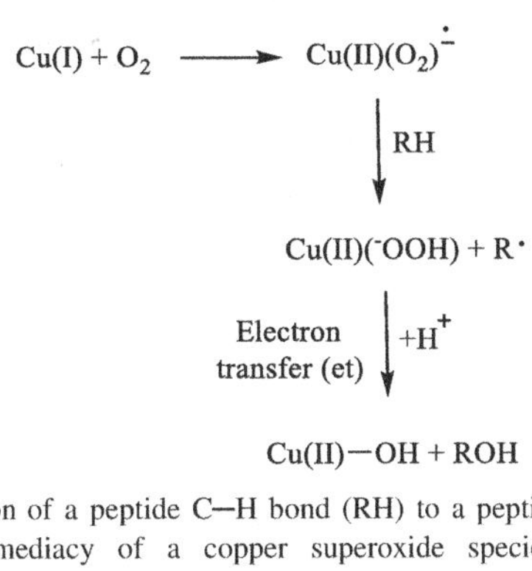

FIGURE 27.4 Conversion of a peptide C—H bond (RH) to a peptide C—OH bond (ROH) by PHM through the intermediacy of a copper superoxide species. *PHM*, Peptidylglycine α-hydroxylating monooxygenase.

bound hydroperoxide. Finally, by one of several possible mechanisms, the O—O bond of the hydroperoxy group is cleaved, and the terminal hydroxyl portion becomes the source of the hydroxyl group in the product alcohol. The cleavage chemistry is associated with electron transfer ("et" from the second copper atom) and protonation.

27.5 Detailed mechanism and the role of the active site residues

Since no specific role has been ascribed to the residues bound to the copper atoms, they are omitted from the mechanisms outlined below in Fig. 27.5. There is general agreement that the reaction is initiated by reduction of both Cu_M(II) and Cu_H(II) to the Cu(I) oxidation state by ascorbic acid (Asc; cf. step a). Oxygen and the peptidylglycine substrate (RH) enter the active site and Cu_M(I) transfers an electron to the oxygen to form superoxide anion and Cu_M(II) (step b). X-ray analysis shows that the oxygen is bound to the copper by a coordinate bond in a head-on manner, and in which most of the electron density resides on the oxygen. In step (c) the superoxide "radical-anion" abstracts a hydrogen atom from the substrate to form the peptidyl glycyl radical and a copper-hydroperoxo species. (Note: this step has recently been disputed; see Wu et al., 2019.) It is at this stage that divergence appears in the current views regarding the mechanistic details. In one approach, the hydroperoxo species is protonated (step d) and water is then lost concurrently with electron transfer from Cu_H to Cu_M (step e; see below regarding the mechanism of the electron transfer between the two copper sites). The two radical sites bond (step f) and hydrolysis lead to the product alcohol (step g).

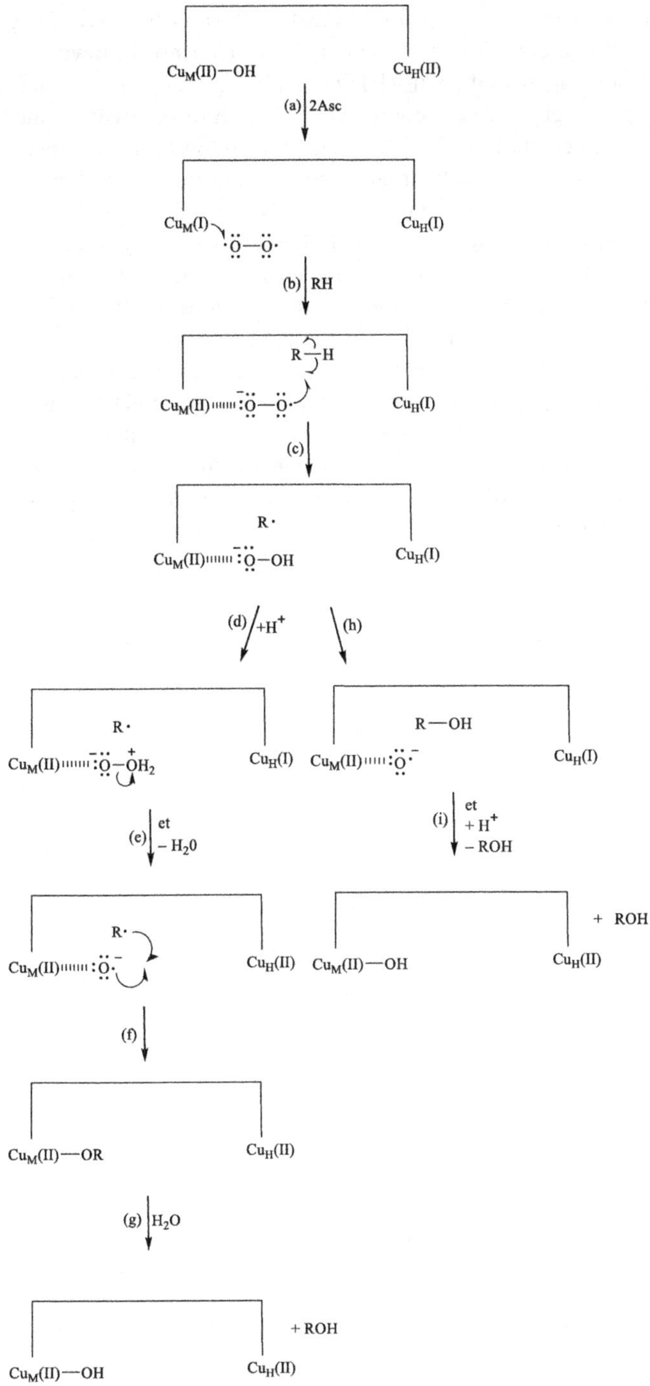

FIGURE 27.5 Detailed mechanism for PHM. *PHM*, Peptidylglycine α-hydroxylating monooxygenase.

An alternative proposal is for a direct transfer of hydroxyl radical to R· (step h; often referred to as a "rebound" mechanism), followed by electron transfer from Cu_H(I) to the Cu_M(II)/O complex, protonation, covalent bonding of the hydroxyl group to the copper, and ejection of the hydroxylated product (step i). An argument has also been made for protonation of the copper-bound superoxide species prior to hydrogen abstraction from the substrate.

There have been few detailed proposals for the mechanism of electron transfer from Cu_H to Cu_M even though this transfer is key to the "uncoupled" nature of the Cu—Cu interaction. Arguments have been made for transfer via the enzyme protein backbone and vis the substrate itself, but the evidence does not appear to support either of these. An argument has been presented that invokes the aqueous solvent that lies between the two copper atoms. This "water-mediated electron transfer" would utilize the highly conserved water network lying between His_{108} on Cu_H and His_{244}. One possible depiction of this chemistry is shown in Fig. 27.6, though electron transfer and bond-breaking/bond-making could well be concerted. Note that the Cu_M species resulting from the oxidation—reduction is that formed in step e of Fig. 27.5.

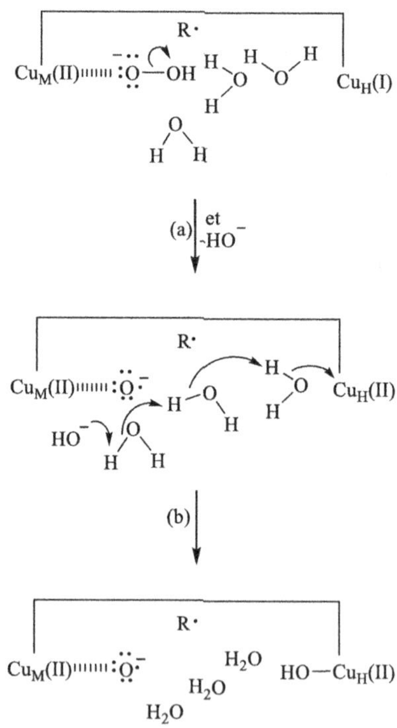

FIGURE 27.6 Proposed role of solvent water molecules in facilitating electron transfer from Cu_H to Cu_M.

Leading references

Wu, P.; Fan, F.; Song, J.; Peng, J.; Liu, J.; Li, C.; Cao, Z.; Wang, B. *J. Am. Chem. Soc.* **141**, 19776–19789 (2019); Solomon, E. I.; Heppner, D. E.; Johnston, E. M.; Ginsbach, J. W.; Cirera, J.; Qayyum, M.; Kieber-Emmons, M. T.; Kjaergaard, C. H.; Hadt, R. G.; Tian, L. *Chem. Rev.* **114**, 3659–3853 (2014); Abad, E.; Rommel, J. B. Kästner, J. *J. Biol. Chem.* **289**, 13726–13728 (2014); Osborne, R. L. and Klinman, J. P. In: Karlin K. D. and Itoh S., Eds; Copper-oxygen chemistry. John Wiley and Sons, Hoboken, NJ, 2011; pp 1–22; Prigge, S. T.; Eipper, B. A.; Mains, R. E.; Amzel, L. M. *Science* **304**, 864–867 (2004); Chen, P. and Solomon, E. I. *J. Am. Chem. Soc.* **126**, 4991–5000 (2004); Evans, J. P.; Ahn, K.; Klinman, J. P. *J. Biol. Chem.* **278**, 49691–49698 (2003); Prigge, S. T.; Kolhekar, A. S.; Mains, R. E.; Amzel, L. M. *Science* **278**, 1300–1305 (1997).

Chapter 28

Phosphatidylinositol-specific phospholipase C

28.1 Phosphatidylinositol-specific phospholipase C

Phosphatidylinositol-specific phospholipase C (EC 3.1.4.11; phosphoinositide phospholipase; PI-PLC) is one of a group of phospholipases that cleaves an ester bond of a phosphorylated glycerol. The PI-PLC family of 13 isozymes cleaves phosphatidylinositol-4,5-bisphosphate (PIP_2) to form a diacylglycerol (*sn*-1,2-diacylglycerol; DAG) and inositol 1,4,5-triphosphate (IP_3) (see Fig. 28.1: R_1 and R_2 are aliphatic chains of fatty acids). All require calcium ion as a cofactor for activity. Note that the *sn* nomenclature refers to "stereospecific numbering" used for Fischer projections of glycerolipids.

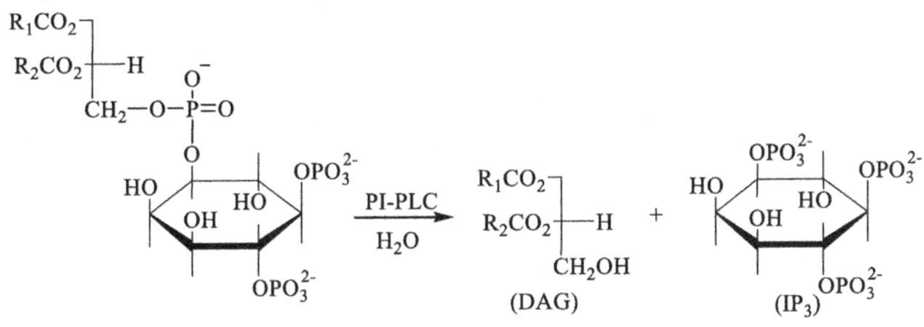

FIGURE 28.1 Cleavage of PIP_2 by PI-PLC. PIP_2, phosphatidylinositol-4,5-bisphosphate; *PI-PLC*, phosphatidylinositol-specific phospholipase C.

28.2 Physiological function

PI-PLC is located primarily in the cytosol and relocates to the plasma membrane upon cellular activation. There it initiates a "phosphoinositide cascade" of secondary messenger-induced hormonal signals by the hydrolysis of PIP_2 to IP_3 and DAG. The two secondary messengers, IP_3 and DAG, activate the release of calcium and protein kinase C, respectively. Several of the PLC subfamilies are activated by the association with G protein (guanine nucleotide-binding protein). PLC-δ1 is distributed widely throughout the body, including the brain, where a possible role in Alzheimer's disease has been raised.

Enzyme Active Sites and their Reaction Mechanisms. DOI: https://doi.org/10.1016/B978-0-12-821067-3.00028-3

28.3 Key structural features

The greatest amount of information about the active site is available for the mammalian isozyme, PLC-δ1, isolated from the rat. Numerous amino acid residues bind to the PIP_2 alcohol and phosphate functionalities, and to the calcium ion, in the active site. These are shown in Fig. 28.2; the figure includes mutations inserted by the authors.

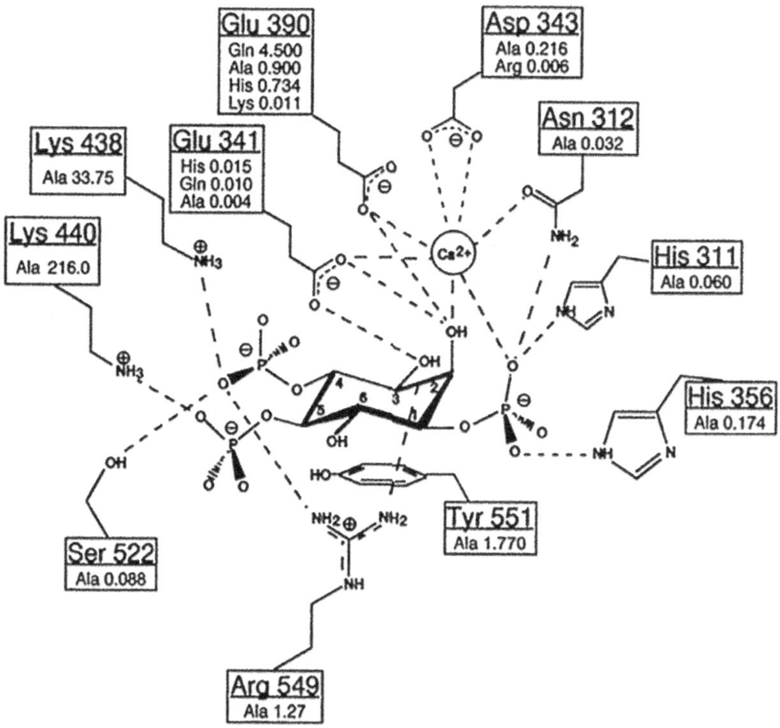

FIGURE 28.2 PIP_2 bound into the active site of PLC-δ1. The native residues are in large type. *PIP_2*, Phosphatidylinositol-4,5-bisphosphate. *This research was originally published in the* Journal of Biological Chemistry. *Ellis, M. V. et al., Catalytic domain of phosphoinositide-specific phospholipase C (PLC).* J. Biol. Chem. *1998. 273, 11650–11659. The American Society for Biochemistry and Molecular Biology.*

28.4 Reaction sequence

The chemistry involved in this reaction is relatively simple—a base-catalyzed hydrolytic cleavage of a phosphate ester (see Fig. 28.3). The

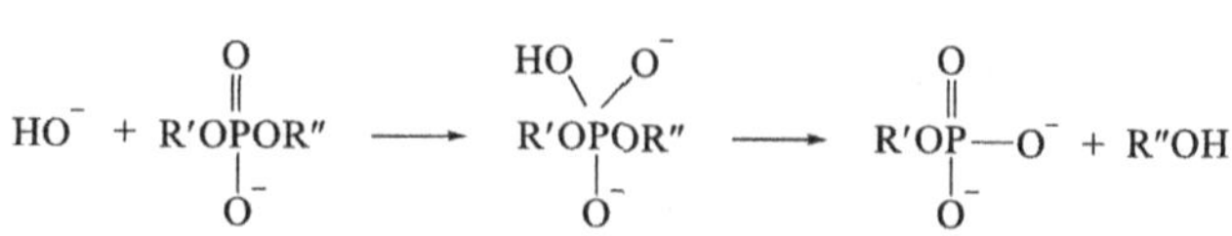

FIGURE 28.3 General reaction sequence for the hydrolysis of a phosphate ester.

addition–elimination mechanism shown here is analogous to that commonly seen for the hydrolysis of carboxylic esters. There are alternative mechanisms for the hydrolysis of a phosphate ester that are analogous to the S_N2 and S_N1 mechanisms for nucleophilic substitution on carbon, but the isolation of a cyclic intermediate (see below) indicates that addition–elimination is occurring here.

28.5 Detailed mechanism and the role of the active site residues

A plausible, but speculative, mechanism is shown in Fig. 28.4. There is broad agreement that His_{311} and His_{356} are key components of the catalytic site, but their exact roles are not well defined. Likewise, the base responsible for the deprotonation of the 2-hydroxy group has not been ascertained with certainty—both of the His residues, as well as Glu_{341}, have been suggested. Glu_{341} is used below because it appears to be better situated for this role. As is seen in other metalloenzymes, the calcium ion increases the acidity of the alcohol thus facilitating its deprotonation. It also helps stabilize the cyclic intermediate. Note also that the hydrolysis by water in steps (d) and (e) is assumed to occur by an addition–elimination mechanism by analogy with steps (a) and (b), but alternatives for step (c) and (d) (see above) are plausible.

In step (a), Glu_{341} abstracts the 2-hydroxyl proton to form an alkoxide anion that attacks the 1-phosphodiester. The newly created oxygen anion is stabilized by coordination with the calcium ion. The product is inositol 1,2-cyclic monophosphate. In step (b), DAG is expelled with the assistance of His_{356}. In step (c), His_{356} facilitates the first step in the hydrolysis of the cyclic phosphate ester, and step (d) completes the hydrolysis.

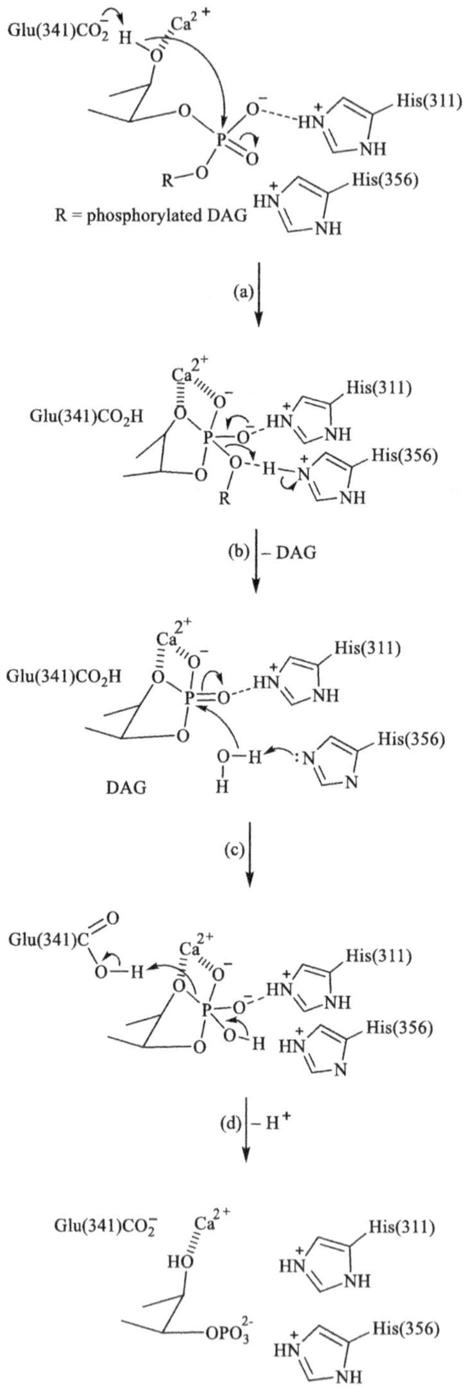

FIGURE 28.4 Detailed mechanism for PLC-δ1 showing the involvement of key amino acids at the active site. *Adapted with permission from Essen et al.* Biochemistry *36, 1704–1718 (1997). Copyright 1997, American Chemical Society.*

Leading references

Roberts, M. F.; Khan, H. M.; Goldstein, R.; Reuter, N.; Gershenson, A. *Chem. Rev.* **118**, 8435–8473 (2018); Moroz, O. V.; Blagova, E.; Lebedev, A. A.; Nørgaard, A.; Segura, D. R.; Blicher, T. H.; Brask, J.; Wilson, K. S. *Acta Cryst.* **D73**, 32–44 (2017); Fukami, K.; Inanobe, S.; Kanemaru, K.; Nakamura, Y. *Prog. Lipid Res.* **49**, 429–437 (2010); Apiyo, D.; Zhao, L.; Tsai, M.-D.; Selby, T. L. *Biochemistry* **44**, 9980–9989 (2005); Ellis, M. V.; James, S. R.; Perisic, O.; Downes, C. P.; Williams, R. L.; Katan, M. *J. Biol. Chem.* **273**, 11650–11659 (1998); Essen, L.-O.; Perisic, O.; Katan, M.; Wu, Y.; Roberts, M. F.; Williams, R. L. *Biochemistry* **36**, 1704–1718 (1997).

Chapter 29

Protein kinase A

29.1 Protein kinase A

Protein kinase A [EC 2.7.11.11; PKA; cyclic adenosine monophosphate (i.e., cyclic AMP)-dependent protein kinase; cAPK] is a "serine protein kinase", an enzyme that targets serine and threonine residues. It constitutes a subgroup of a superfamily of over 2000 vertebrate protein kinases. The enzyme depends on the cellular level of cAMP (see Fig. 29.1), which is formed from ATP by the enzyme, adenylyl cyclase. Four cAMP molecules are needed to activate one PKA, which they do by binding to a pair of regulatory PKA subunits. The bound subunits then detach, leaving two activated catalytic subunits, labeled PKAc. This catalytic subunit contains only 350 AA residues and has thus been the subject of extensive experimental and theoretical studies.

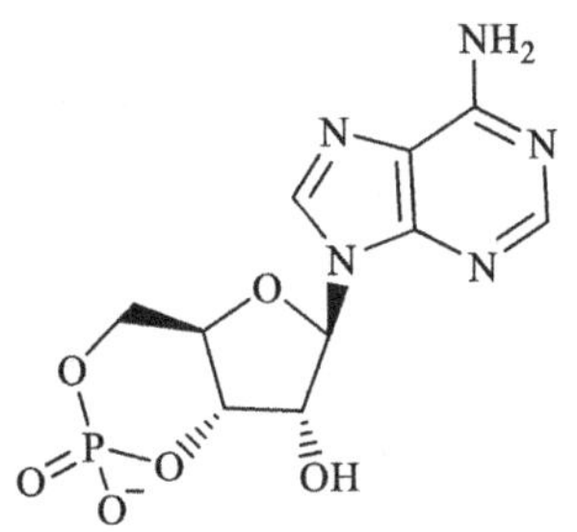

FIGURE 29.1 Cyclic adenosine monophosphate.

The role of a protein kinase is to transfer the γ-phosphate of magnesium adenosine triphosphate (MgATP) to other proteins. PKA phosphorylates serine and threonine residues; the tyrosine hydroxyl group is phosphorylated by a different enzyme. The overall reaction is shown in Fig. 29.2. The rates of kinase-catalyzed phosphorylation have been estimated to be 10^{14} times greater than for comparable, uncatalyzed, reactions.

$$\text{ADP—OPO}_3^{2-} + \text{ROH} \longrightarrow \text{ADP—O}^- + \text{ROPO}_3^{2-} + \text{H}^+$$

FIGURE 29.2 The overall reaction catalyzed by protein kinase A. ROH represents a serine or threonine residue.

Enzyme Active Sites and their Reaction Mechanisms. DOI: https://doi.org/10.1016/B978-0-12-821067-3.00029-5

Fig. 29.3 is a table that defines the PKA target "consensus peptide recognition" sequence and the two polypeptides that have most often been used as inhibitors for structural studies. The "P-site" represents the serine. Adjacent substrate residues consist of a pair of arginines C-terminal to the P-site at P^{-2} and P^{-3}, and a small AA at P^{-1}. A hydrophobic group N-terminal to the P-site is at P^{+1}. The inhibitor polypeptides are "Kemptide" and PKI(5-24); each contains an alanine residue in place of the serine. Fig. 29.4 shows the interactions of the P^{+1} to P^{-3} residues with AAs within the catalytic site. Of note, in this figure are the two P^{-2} and P^{-3} Arg residues interactions with four Glu units, and the location of the P^{+1} residue within a hydrophobic pocket.

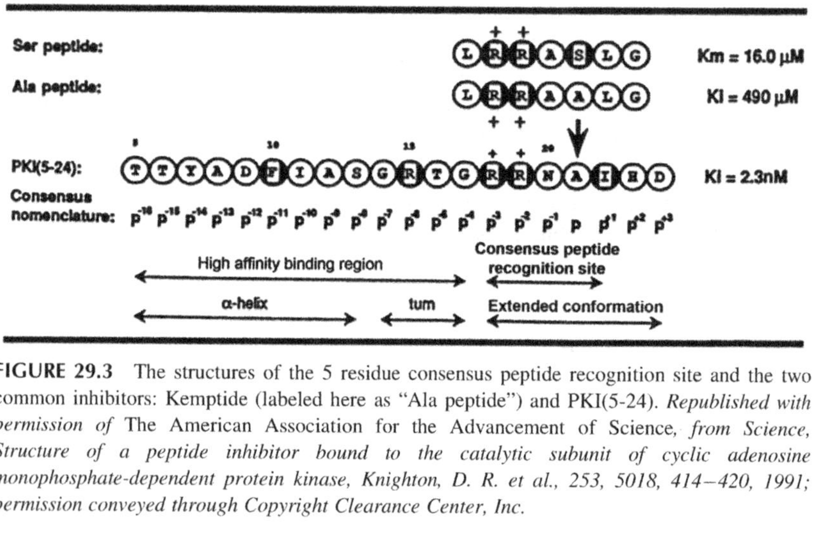

FIGURE 29.3 The structures of the 5 residue consensus peptide recognition site and the two common inhibitors: Kemptide (labeled here as "Ala peptide") and PKI(5-24). *Republished with permission of* The American Association for the Advancement of Science, *from Science, Structure of a peptide inhibitor bound to the catalytic subunit of cyclic adenosine monophosphate-dependent protein kinase, Knighton, D. R. et al., 253, 5018, 414–420, 1991; permission conveyed through Copyright Clearance Center, Inc.*

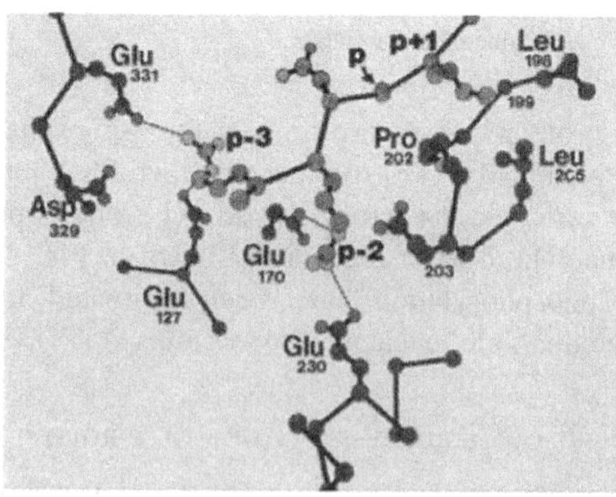

FIGURE 29.4 AA interactions at the recognition site. *AAAS, 1991; See Fig. 29.3.*

29.2 Physiological function

Phosphorylation activates and regulates numerous critical cellular enzymes including those involved in cell signal transduction, cellular division and cellular differentiation, and carbohydrate and lipid metabolism.

29.3 Key structural features

The C subunit is made up of two lobes, one large (residues 128–300) and one small (residues 15–127). The small lobe contains the binding site for the MgATP, whereas the large lobe contains most of the residues associated the docking and catalysis of the target protein substrate. The binding of the MgATP and the substrate occurs independent of one another. The active site is situated in a cleft between the two lobes. This is seen in Fig. 29.5; here, as in most structural studies, an inhibitor peptide, PKI(5-24), is bound to the enzyme.

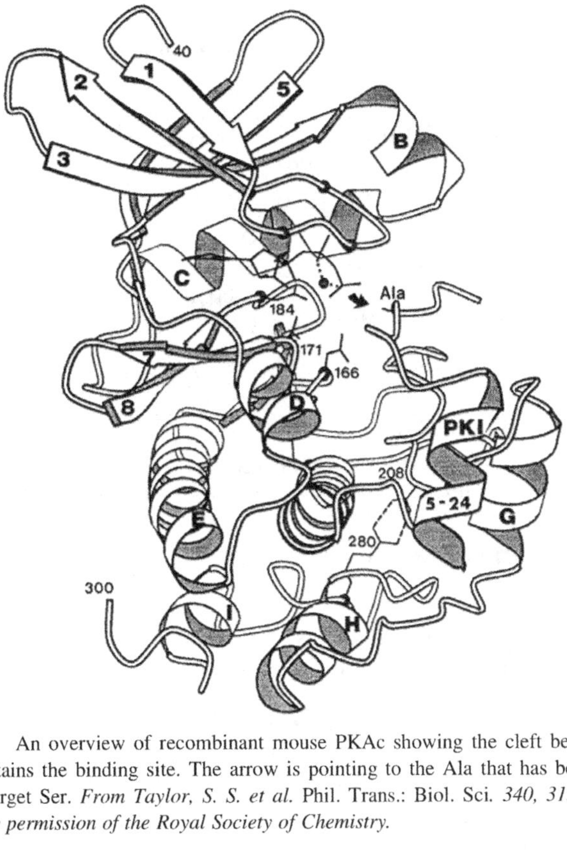

FIGURE 29.5 An overview of recombinant mouse PKAc showing the cleft between the two lobes that contains the binding site. The arrow is pointing to the Ala that has been inserted in place of the target Ser. *From Taylor, S. S. et al.* Phil. Trans.: Biol. Sci. *340, 315–324 (1993). Reproduced by permission of the Royal Society of Chemistry.*

Several regions have been identified and studied extensively. These include an ATP "nucleotide binding loop" containing a "glycine-rich" sequence of Gly_{50}, Thr_{51}, Gly_{52}, Ser_{53}, Phe_{54}, and Gly_{55}, a "catalytic loop" of Arg_{165}, Asp_{166}, Leu_{167}, Lys_{168}, Pro_{169}, and Asn_{170}, and an "activation loop" that includes a threonine residue (Thr_{197}) that must be phosphorylated in order for the protein to adopt the necessary conformation for activity. Additional studies have emphasized the interactions of the groups of residues adjacent to the "P-site" (see above) with those in the catalytic loop. For example, the "P + 1" loop consists of residues 198–205. Leu_{167} and Asp_{166} of the catalytic loop interact with Tyr_{204} and Thr_{201}, respectively.

A representation of the active site with several key residues is shown in Fig. 29.6. Asp_{166}, Glu_{91}, Lys_{72}, and Lys_{168} play significant roles in organizing the reactants and stabilizing the transition state (TS). Two magnesium ions (typically labeled Mg_1 and Mg_2) appear at the active site, and they are depicted with their associated bound residues, Asp_{184} and Asn_{171}. Mg_1 has been labeled the "primary" magnesium; it helps position the gamma phosphate for the nucleophilic attack.

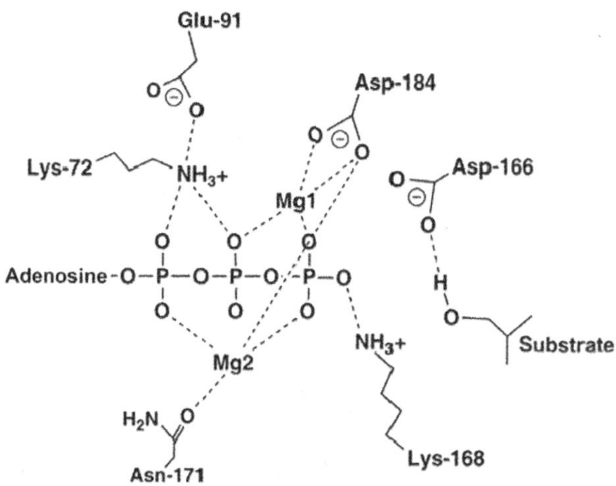

FIGURE 29.6 The active site of PKA. *PKA*, Protein kinase A. *Adapted with permission from Adams, J. A.* Chem. Rev. *101, 2271–2290 (2001). Copyright 2001, American Chemical Society.*

29.4 Reaction sequence

The chemistry is basically that of a transesterification (see Fig. 29.7). Though the reaction in this figure is shown as base-catalyzed, the role of a base in the PKA reaction is still debated (see below).

$$ROH + B^- \rightleftharpoons RO^- + BH$$

$$ADP{-}OPO_3^{2-} + RO^- \longrightarrow ADP{-}O^- + ROPO_3^{2-}$$

FIGURE 29.7 General reaction sequence for the base-catalyzed phosphorylation of an alcohol using a phosphate ester.

29.5 Detailed mechanism and the role of the active site residues

A more detailed description of the mechanism, as it involves several (but not all) key residues in the active site, is shown in Fig. 29.8. Isotopic labeling studies have confirmed that the reaction proceeds with inversion at the phosphorous atom, and there is no evidence for an enzyme-bound intermediate (i.e., the reaction is "concerted"). Step (a) shows the formation of the TS resulting from the serine hydroxyl group initiating attack on the gamma phosphate. This nucleophilic attack is aided by hydrogen bonding to, but not

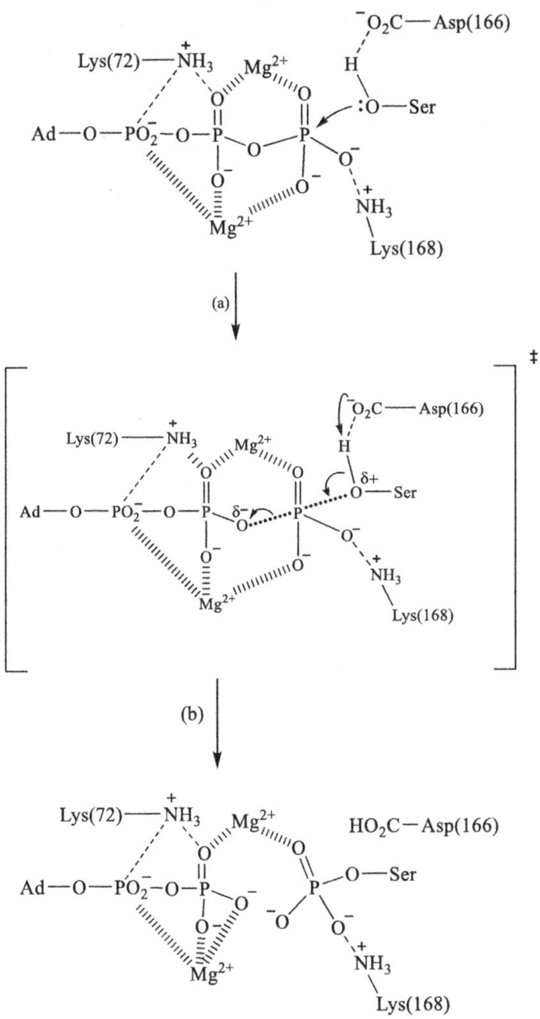

FIGURE 29.8 Detailed mechanism for PKA phosphorylation of a serine residue showing the involvement of key amino acids at the active site. *PKA*, Protein kinase A.

full deprotonation by, Asp_{166}. Partial cleavage of the P—O bond occurs concomitantly with this attack, but as noted below, the breakage of the P—O bond in the TS is much farther along than is bond-making with the serine. The reaction concludes in step (b) with complete dissociation of the magnesium-chelated Adp "leaving group," phosphorylation of the serine and proton transfer to the aspartate.

There have been two primary items of dispute regarding this mechanism. One is the relative timing of the phosphate–phosphate O—P bond-breaking relative to the serine O—P bond-making. When the phosphate/phosphate bond breaks well before the new serine (or threonine) phosphate bond is formed, the corresponding TS is called "dissociative." Conversely, if the serine begins bonding to the gamma phosphate phosphorous atom well before bond-breaking occurs, the TS is called "associative." Evidence strongly favors the dissociative process. The second area of dispute has been the role of the highly conserved Asp_{166} It may act as a base, deprotonating the serine before nucleophilic attack sets in. Alternatively, it may act more as a directing agent, lining up the OH group for optimal attack on the phosphorous through hydrogen bonding, but not actually removing the proton until late in the reaction. The current view favors this latter function and the mechanism in Fig. 29.8 is depicted accordingly.

Leading references

Solorza, J.; Recabarren, R.; Alzate-Morales, J. *J. Chem. Inf. Model.* **60**, 898–914 (2020); Gerlits, O.; Tian, J.; Das, A.; Langan, P.; Heller, W. T.; Kovalevsky, A. *J. Biol. Chem.* **290**, 15538–15548 (2015); Yang, J.; Ten Eyck, L. F.; Xuong, N.-H.; Taylor, S. S. *J. Mol. Biol.* **336**, 473–487 (2014); Endicott, J. A.; Noble, M. E. M.; Johnson, L. N. *Annu. Rev. Biochem.* **81**, 587–613 (2012); Yang, J.; TenEyck, L. F.; Xuong, N.-H.; Taylor, S. S. *J. Mol. Biol.* **336**, 473–487 (2004); Johnson, D. A.; Akamine, P.; Radzio-Andzelm, E.; Madhusudan.; Taylor, S. S. *Chem. Rev.* **101**, 2243–2270 (2001); Adams, J. A. *Chem. Rev.* **101**, 2271–2290 (2001); Taylor, S. S.; Zheng, J.; Radzioandzelm, E.; Knighton, D. R.; Ten Eyck, L. F.; Sowadski, J. K.; Herberg, F. W.; Yonemoto. *Phil. Trans. Biol. Sci.* **340**, 315–324 (1993); Knighton, D. R.; Zheng, J. H.; Ten Eyck, L. F.; Xuong, N. H.; Taylor, S. S.; Sowadski, J. M. *Science* **253**, 407–414 (1991); Knighton, D. R.; Zheng, J.; Ten Eyck, L. F.; Xuong, N.-H.; Taylor, S. S.; Sowadski, J. M. *Science* **253**, 414–420 (1991).

Chapter 30

Pyruvate carboxylase

30.1 Pyruvate carboxylase

Pyruvate carboxylase (EC 6.4.1.1; PC) is a "Class I," biotin-dependent, mitochondrial protein that catalyzes the conversion of pyruvate to oxaloacetate. The structure of biotin is shown in Fig. 30.1. The overall reaction is shown in Fig. 30.2. Other members of this enzyme family include acetyl-CoA carboxylase, propionyl-CoA carboxylase, 3-methylcrotonyl-CoA carboxylase, geranyl-CoA carboxylase, and urea carboxylase. These share the use of adenosine triphosphate (ATP; Fig. 30.3), bicarbonate, and divalent metal

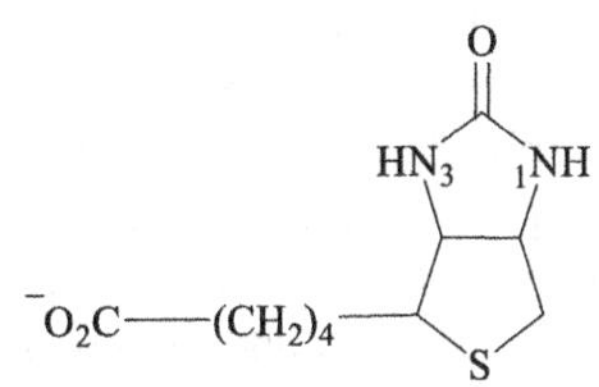

FIGURE 30.1 The chemical structure of biotin.

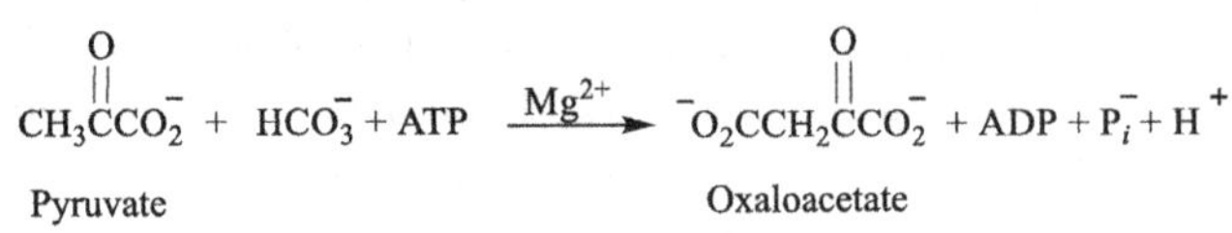

FIGURE 30.2 Overall reaction catalyzed by pyruvate carboxylase.

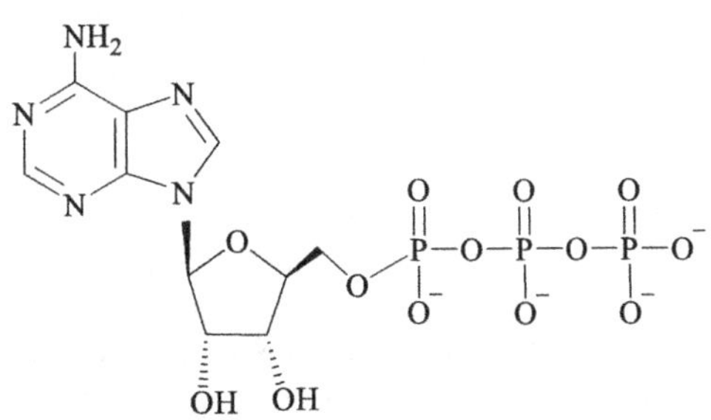

FIGURE 30.3 Chemical structure of adenosine triphosphate.

Enzyme Active Sites and their Reaction Mechanisms. DOI: https://doi.org/10.1016/B978-0-12-821067-3.00030-1

ions as enzyme cofactors (Mg^{2+} for *Rhizobium etli*; Mn^{2+} for vertebrates). Mammalian PC also employs acetyl-CoA as an allosteric activator (see Chapter 15: Dihydrolipoyl Transacetylase for the structure of acetyl-CoA).

30.2 Physiological function

PC is found in most living organisms. Oxaloacetate is an intermediate in the tricarboxylic acid (i.e., Krebs) cycle. As a biosynthetic source of oxaloacetate, PC is involved in gluconeogenesis, and lipogenesis. The formation of oxaloacetate also leads to higher levels of NADPH, which in turn enhances insulin secretion from the pancreatic islets.

30.3 Key structural features

Vertebrate PC is a multifunctional enzyme typically consisting of four identical subunits (i.e., is a tetramer). Each subunit contains a single polypeptide chain, which embodies three functional domains, plus an allosteric domain (also called the pyruvate tetramerization domain) that binds acetyl-CoA. Only the tetrameric form of the enzyme is active. The three functional domains are (1) biotin carboxylase (BC), (2) carboxytransferase (CT), and (3) biotin-carboxyl carrier protein (BCCP). In summary, biotin is carboxylated in the BC domain, the carboxybiotin is tethered to the BCCP through its valerate side chain to a lysine residue, and the bound biotin transferred to the CT domain on an opposing (partner) polypeptide chain within the tetramer. One role of the acetyl-CoA is to induce a conformational change that brings the BC and CT domains on separate chains closer to one another. Thus the allosteric domain is located at the interfaces of the BC and CT, and CT and BCCP, domains. Nevertheless, the BCCP domain "swings" between the BC and CT domains using a c.20 residue "linker arm."

The allosteric binding site for *R. etli* (RePC) for ethyl-CoA, an oft used analog of acetyl-CoA, shows the principle binding residues to be Arg_{427}, Arg_{429}, Arg_{469}, Asp_{471}, and Arg_{472}. The pocket for ATP within the BC catalytic active site is characteristic of a family of "ATP-grasp enzymes." The key residues (determined for RePC using the nonhydrolysable sulfur analog, adenosine 5′*O*-(3-thiophosphate; ATP-γ-S) are shown in Fig. 30.4.

The several key residues in the binding site for biotin in the RePC BC domain are shown in Fig. 30.5 (note again the use of ATP-γ-S). The biotin is attached to the enzyme via bonding with a lysine residue.

An outline of primary residue/substrate interactions within the human CT domain is shown in Fig. 30.6. The Gln_{575} and Arg_{644} diyad is heavily conserved within the PC family, as is the Ser_{911}. The divalent metal atom is Zn^{2+} in yeast and bacteria. The Lys_{741} is carbamylated (Fig. 30.7).

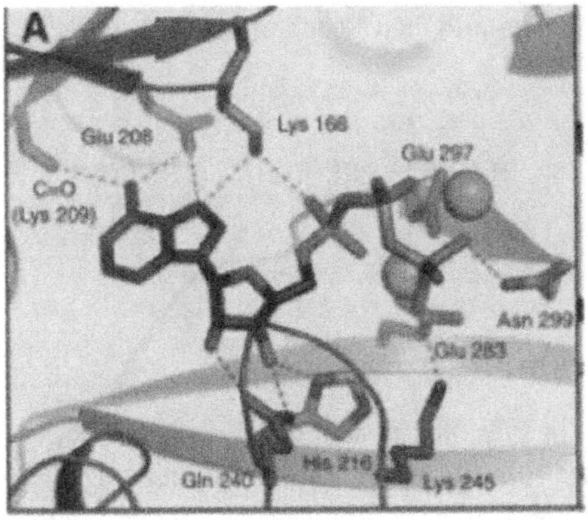

FIGURE 30.4 Key residues involved in the binding site for ATP. ATP, Adenosine triphosphate. *(Republished with permission of the American Academy of Arts and Sciences, from Domain architecture of pyruvate carboxylase, a biotin-dependent multifunctional enzyme, St. Maurice, M. et al.* Science ***317*** *(5841), 1076–1079 2007; permission conveyed through Copyright Clearance Center, Inc.).*

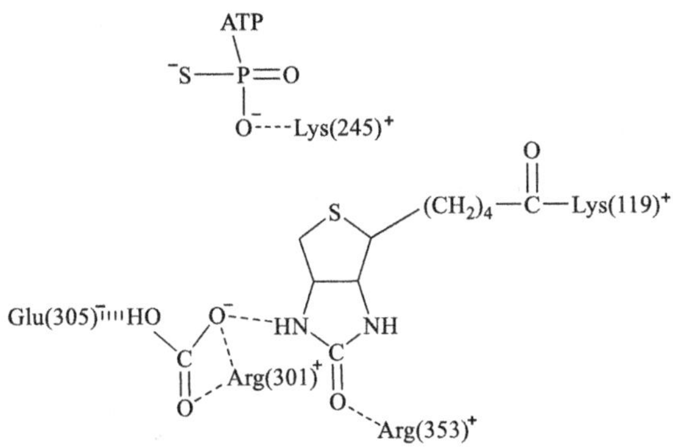

FIGURE 30.5 Key residues involved in the binding site for biotin.

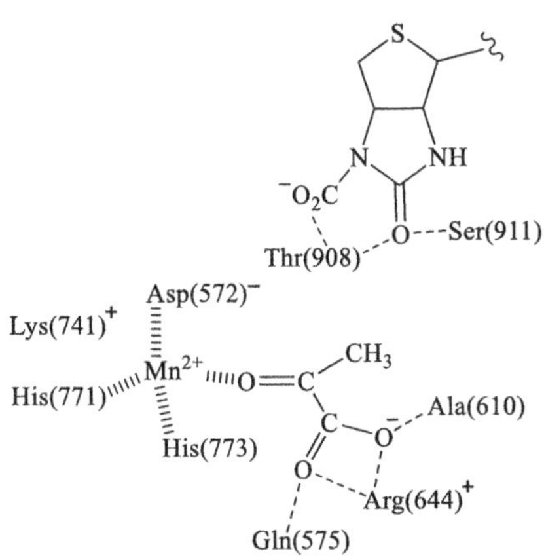

FIGURE 30.6 Key residues involved in the active site of the human CT domain. *(Adapted by permission from Springer: Nat. Struct. Mol. Biol. Crystal structures of human and* Saphylococcus aureus *pyrivate carboxylase and molecular insights into the carboxyl transfer reaction, Xiang, S. and Tong, L. Copyright 2008).*

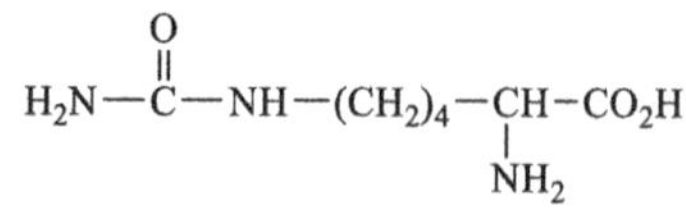

FIGURE 30.7 The chemical structure of carbamylated lysine.

30.4 Reaction sequence

The reaction shown in Fig. 30.2 occurs in a two-step process (Fig. 30.8). The first reaction (a) is the carboxylation of biotin at N_1 utilizing the bicarbonate anion to ADP-Mg^{2+} (*R. etli*). Note that the presence of a second metal cation serves to catalyze this reaction, presumably by helping to optimally orient the ATP. The second step (b) involves the transfer of the carboxyl group from biotin to pyruvate to form oxaloacetate and the reformation of the biotin catalyst. As noted above, the carboxylated biotin must be transferred from the BC domain to the CT domain in order for step (2) to occur. It is tethered to the BCCP during that transition.

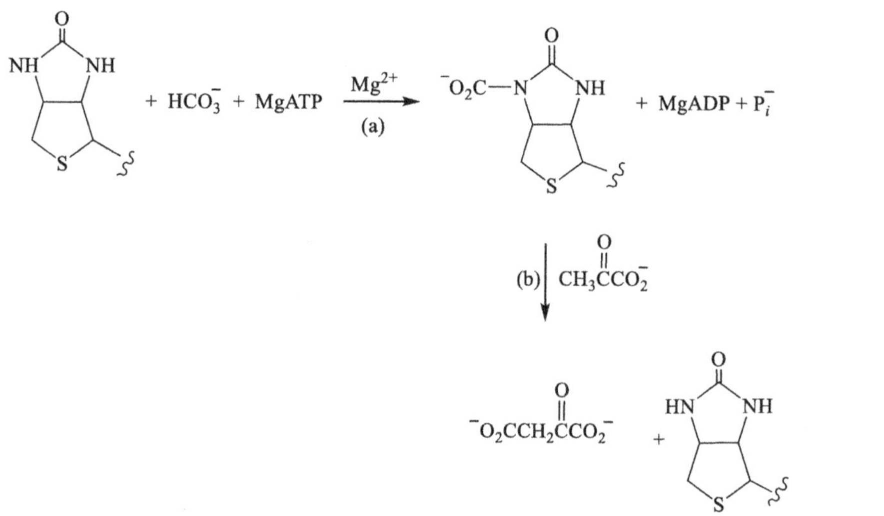

FIGURE 30.8 Two-step reaction sequence for pyruvate carboxylase.

30.5 Detailed mechanism and the role of active site residues

The first portion of the carboxyl transfer sequence is generally agreed to be the carboxylation of biotin (see Fig. 30.9). In the first step, the Glu_{305} carboxylate ion (using *R. etli* numbering) deprotonates the bicarbonate ion. It has been proposed that this deprotonation is actually initiated by a sequential set of proton transfers wherein His_{216} deprotonates Glu_{218}, which in turn deprotonates Glu_{305}. The carbonate species thus generated attacks the ATP. Lys_{245} helps orient the ATP by H-bonding to the terminal phosphate. (Note that the nature of the divalent metal ion varies with the species).

Typically, the "ester interchange" reaction would proceed by an addition–elimination sequence leading to a transient carboxyphosphate species (steps (a) and (b)). The carboxyphosphate decomposes into carbon dioxide, inorganic phosphate, and Adp (step (c)). The phosphate ion is then proposed to deprotonate biotin to form a biotin enolate that is stabilized by a nearby Arg_{353} (step (d); the relative pKa's of the two species would normally seem to make this proposal problematical). This anion attacks the carbon dioxide to form the carboxylated biotin intermediate (step (e)). An alternative process has also been considered, wherein the bicarbonate attacks the ATP first and then deprotonates before fragmenting to carbon dioxide.

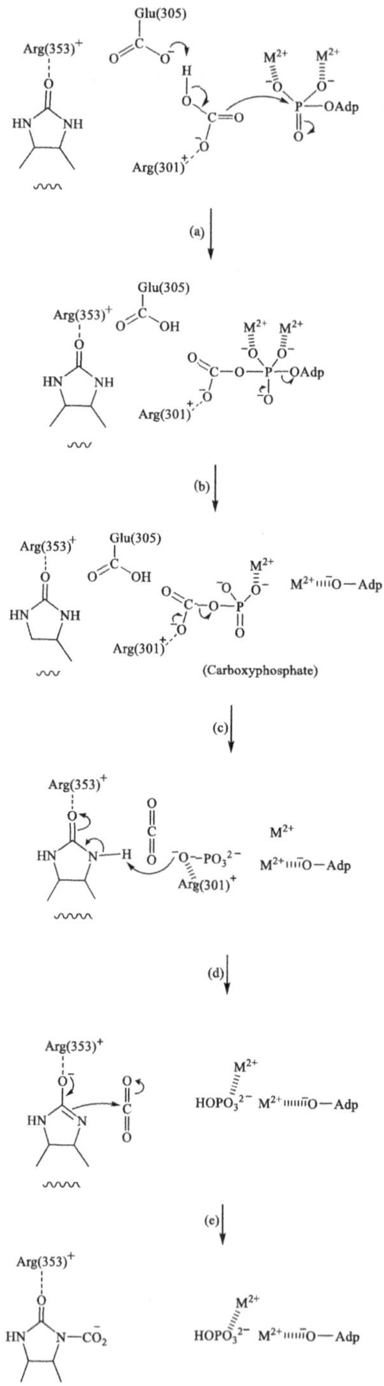

FIGURE 30.9 Mechanism for the carboxylation of biotin. *(Adapted with permission from Zeczycki, T. N. et al.* Biochemistry ***50****, 9724–9737 (2011); Copyright 2011, American Chemical Society).*

The carboxyl transfer chemistry is outlined in Fig. 30.10. Note this occurs upon translocation of the (Lys_{1119}-tethered) carboxybiotin from the BC domain to the CT domain. The carboxybiotin decarboxylates to CO_2 and the biotin enolate (step (a)). It has been proposed that Ser_{885} and

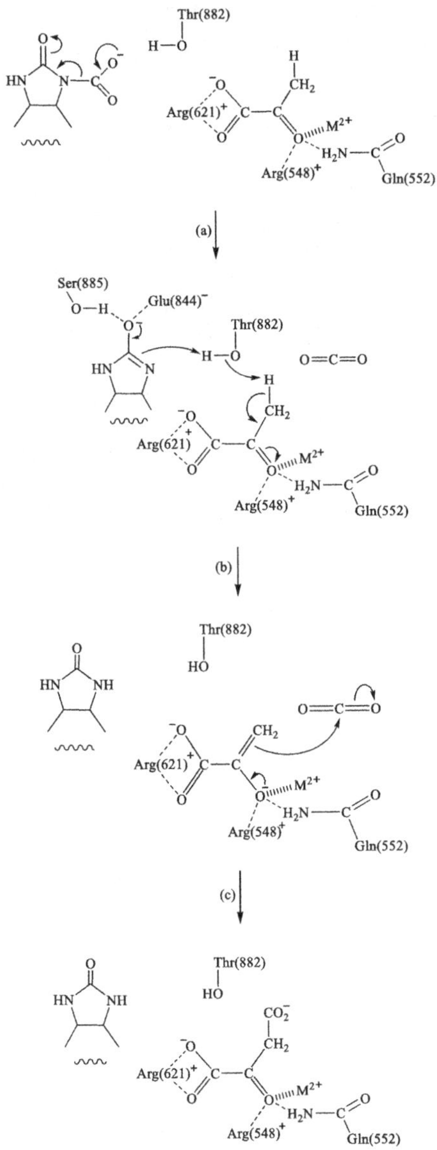

FIGURE 30.10 Mechanism of the carboxyl transfer chemistry. *(From Lietzan, A. D. et al. The role of biotin and oxamate in the carboxyltransferase reaction of pyruvate carboxylase,* Arch. Biochem. Biophys. ***562****, 70–79 (2014)).*

Glu_{844} likewise stabilize the negatively charged biotin enolate. The enolate, through the intermediacy of Thr_{882}, deprotonates pyruvate to form the pyruvate enolate (step (b)). Computation suggests that a threonine alkoxide anion may be an intermediate in this double-proton transfer sequence. The pyruvate enolate then attacks the CO_2 to form oxaloacetate (step (c)). Note the stabilization of the pyruvate provided by Arg_{621}, Gln_{552}, and the divalent metal ion.

Leading references

Sheng, X.; Hou, Q.; Liu, Y. *Theor. Chem. Acc.* **138**, 17 (2019); Valle, M. "Pyruvate Carboxylase, Structure and Function". In: Harris J. and Marles-Wright J., Eds; Macromolecular Protein Complexes. Subcellular Biochemistry, vol 83. Springer, Cham, 2017; Sirithanakorn, C.; Jitrapakdee, S.; Attwood, P. V. *Biochemistry* **55**, 4220–4228 (2016); Westerhold, L. E.; Adams, S. L.; Bergman, H. L.; Zeczycki, T. N. *Biochemistry* **55**, 3447–3460 (2016); Menefee, A. L. and Zeczycki, T. N. *FEBS J.* **281**, 1333–1354 (2014); Adina-Zada, A.; Jitrapakdee, S.; Wallace, J. C.; Attwood, P. V. *Biochemistry* **53**, 1051–1058 (2014); Lietzan, A. D. and St. Maurice, M. *J. Biol. Chem.* **288**, 19915–19925 (2013); Tong, L. *Cell. Mol. Life Sci.* **70**, 864–891 (2013); Zecycki, T. N.; Menefee, A. L.; Abdussalam, A.-Z.; Jitrapakdee, S.; Surinya, K. H.; Wallace, J. C.; Attwood, P. V.; St. Maurice, M.; Cleland, W. W. *Biochemistry* **50**, 9724–9737 (2011); Chou, C.-Y.; Yu, L. P. C.; Tong, L. *J. Biol. Chem.* **284**, 11690–11697 (2009); Xiang, S. and Tong, L. *Nat. Struct. Mol. Biol.* **15**, 295–302 (2008); St. Maurice, M.; Reinhardt, L.; Surinya, K. H.; Attwood, P. V.; Wallace, J. C.; Cleland, W. W.; Rayment, I. *Science* **317**, 1076–1079 (2007).

Chapter 31

Pyruvate dehydrogenase

31.1 Pyruvate dehydrogenase

Pyruvate dehydrogenase (EC 1.2.4.1; E1p) is one enzymatic component of the mitochondrial-based pyruvate dehydrogenase multi-enzyme complex (PDHC; PDC; PDH). This is an associated set of three enzymes that ultimately converts pyruvate to acetyl coenzyme A (acetyl-CoA); see Fig. 31.1. Other components of the PDHC are dihydrolipoyl transacetylase (E2p, EC 2.3.1.12) and dihydrolipoyl dehydrogenase (E3p, EC 1.8.1.4). E2p catalyzes the transfer of the acetyl group to CoA to give acetyl CoA and dihydrolipoamide-E2p. E3p oxidizes the dihydrolipoamide to lipoamide, which is then available to E2p to continue the catalytic cycle. See elsewhere in this book for discussions of the chemistry catalyzed by E2p and E3p.

$$\underset{\text{(Pyruvate)}}{CH_3\overset{O}{\overset{\|}{C}}CO_2^-} + CoASH + NAD^+ \longrightarrow \underset{\text{(Acetyl CoA)}}{CH_3\overset{O}{\overset{\|}{C}}SCoA} + CO_2 + NADH$$

FIGURE 31.1 The overall chemistry catalyzed by the pyruvate dehydrogenase multienzyme complex.

E1p is a tetramer consisting of two pairs of subunits (α and β) with two separate active sites. There is communication between the two sites through proton transfer reactions, but only one site is active at any given time. Thiamine diphosphate (TPP; ThDP; thiamine pyrophosphate; Fig. 31.2) is a critical cofactor for E1, as is a magnesium ion. The chemistry catalyzed by E1p is shown in Fig. 31.3.

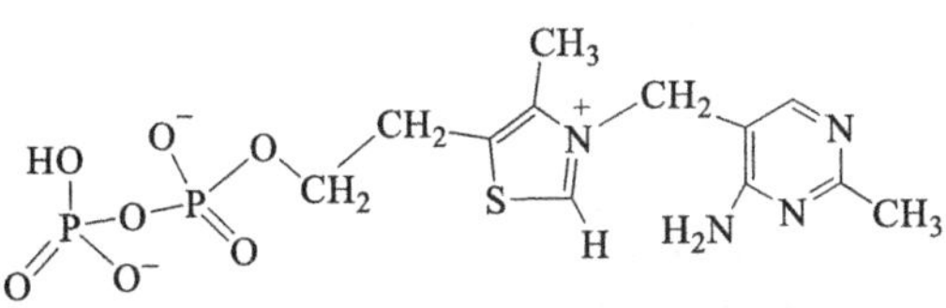

FIGURE 31.2 Thiamine pyrophosphate.

Enzyme Active Sites and their Reaction Mechanisms. DOI: https://doi.org/10.1016/B978-0-12-821067-3.00031-3

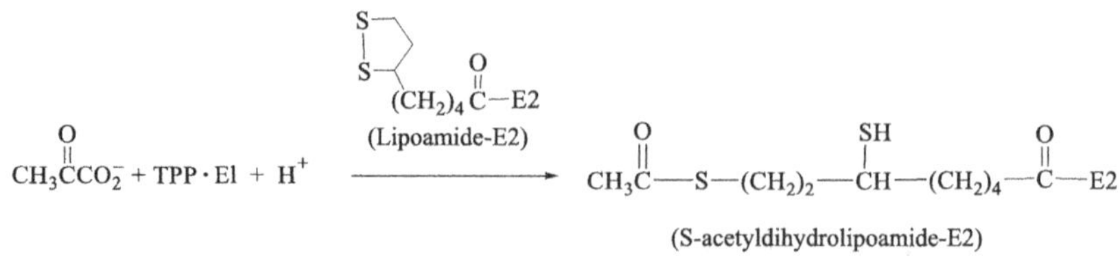

FIGURE 31.3 The chemistry catalyzed by E1p.

31.2 Physiological function

Acetyl-CoA is utilized in the citric acid cycle. The PDHC serves to link glycolysis to this key component of cellular respiration.

31.3 Key structural features

Fig. 31.4 shows the binding pocket for TPP in human pyruvate dehydrogenase. The active site that binds a magnesium ion and TPP lies at the interface between a pair of alpha and beta subunits. Residues from both subunits are involved in the binding interactions. The magnesium ion is held in place by Asn_{195}, Tyr_{198}, two TPP phosphate oxygen atoms, and a water molecule. Gly_{168}, Arg_{90}, Tyr_{89} Gly_{136}, Glu_{59}, Ile_{57}, and Phe_{85} all further stabilize the bound TPP through hydrogen bonding, hydrophobic and pi-pi stacking interactions. One critical consequence of these interactions is the bending of TPP into a V conformation such that N4′ is brought close to the C2-H (see mechanism below). Though some distance from the active site, $Tryp_{135}$, Pro_{188}, Met_{181}, His_{15}, and Arg_{349} all affect enzymatic activity through their

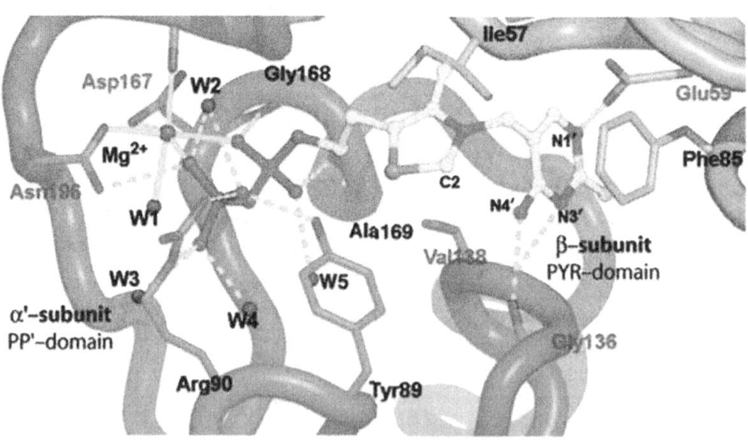

FIGURE 31.4 The key enzyme residues forming the human TPP binding pocket. *TPP*, Thiamine diphosphate. *This research was originally published in the Journal of Biological Chemistry. Ciszak et al.* J. Biol. Chem. ***276****, 21240–21246 (2003). Copyright 2003 The American Society for Biochemistry and Molecular Biology.*

role in maintaining the structural integrity of E1p and/or the requisite conformations of the binding sites. Met_{81} and Met_{200} control access to the TPP binding site. Enzymes utilizing TPP as a cofactor all share a requirement for a Glu residue to initiate reaction by serving as a proton donor.

31.4 Reaction sequence

In the first step, E1p converts pyruvate to enzyme-bound hydroxyethylidene-TPP (HETTP). This is followed by the E1p-catalyzed transfer of an acetyl moiety to "lipoamide-E2p" (lipoic acid covalently bound via an amide linkage to a lysine unit of E2p) to form S-acetyldihydrolipoamide-E2p and reactivated E1p (see Fig. 31.5).

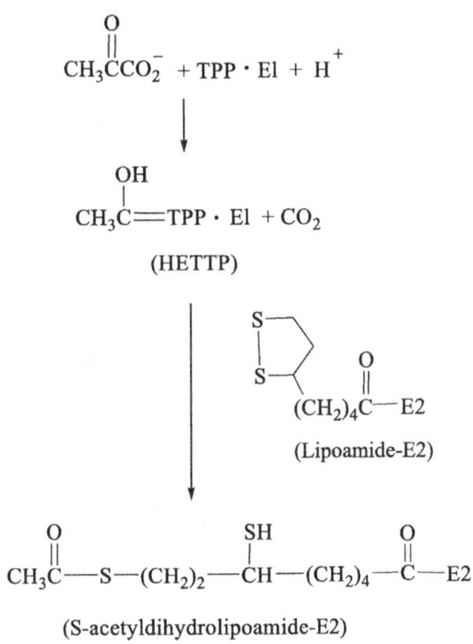

FIGURE 31.5 Reaction sequence for the conversion of pyruvate to S-acetyldihydrolipoamide-E2.

31.5 Detailed mechanism and role of the active site residues

This is shown in Fig. 31.6. The most extensive mechanistic studies have involved the PDH multienzyme complex isolated from *Bacillus stearothermophilus* and the numbering is from that source. The reaction is initiated by Glu_{59} facilitating the tautomerization of TPP in step (a) by protonating N1′. (Though the state of protonation of this acid has been debated, X-ray images indicate it is protonated at the outset; cf. Kluger and Tittmann, 2008). The Glu_{59} carboxylate anion thus formed then deprotonates N1′ with concomitant

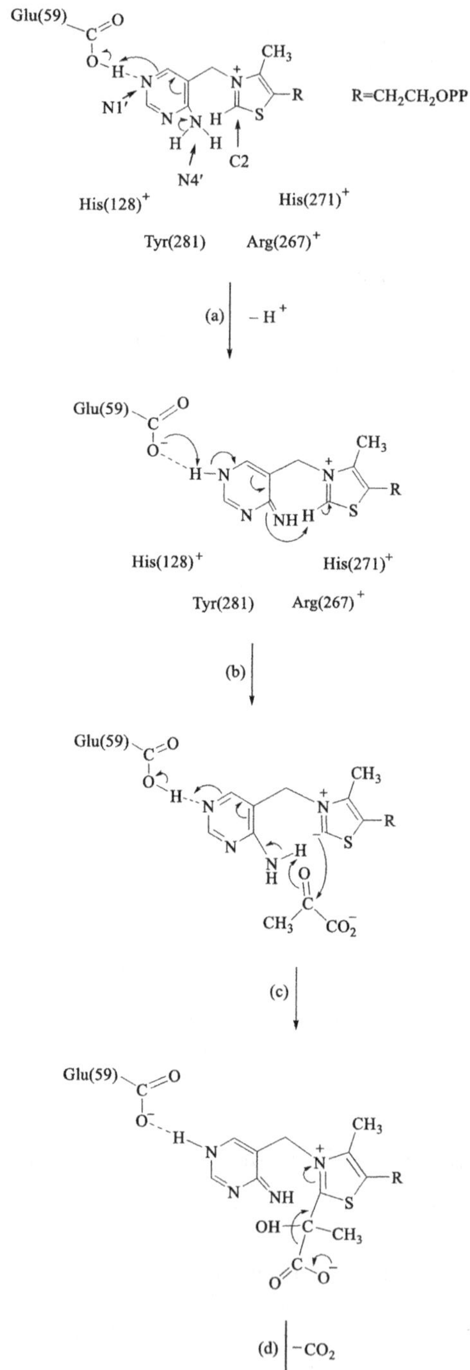

FIGURE 31.6 Mechanism with Glu_{59} catalysis. *Adapted with permission from Fries et al.,* Biochemistry *42, 6996–7002 (2003); Copyright 2003 American Chemical Society.*

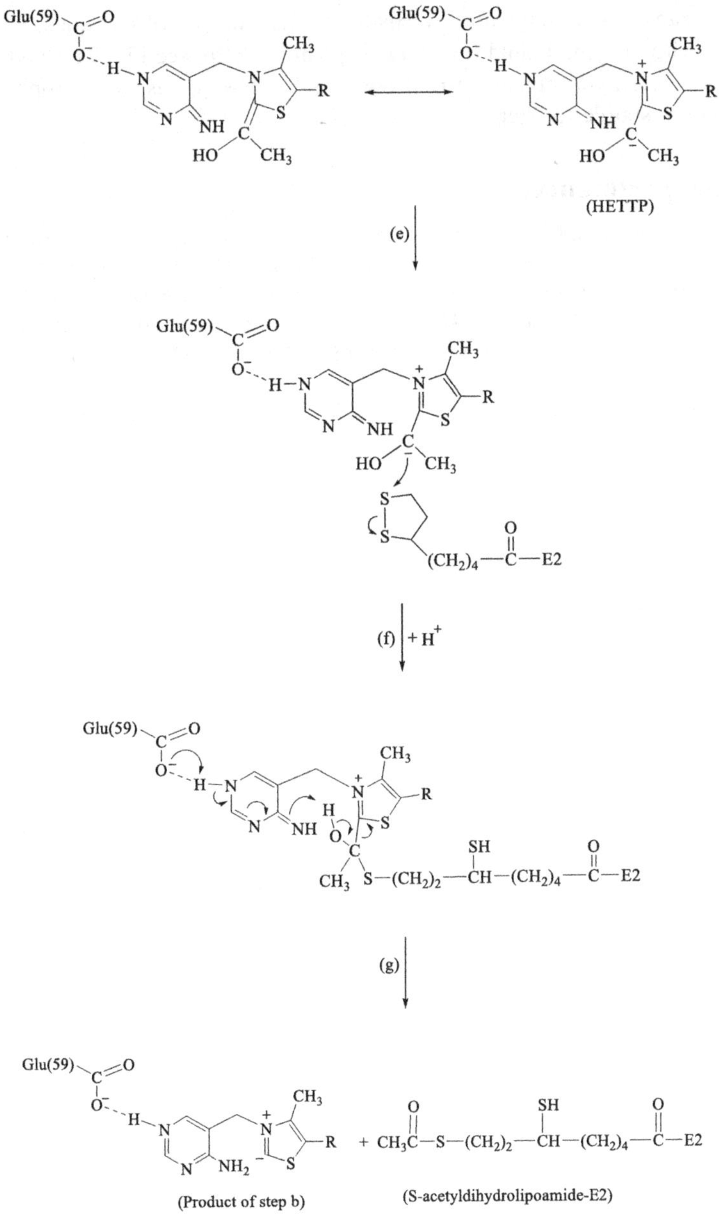

FIGURE 31.6 (Continued)

deprotonation at C2 to form an ylide (step b). Nucleophilic attack on pyruvate by the C2 carbanion leads to HETTP (steps c and d), which is a resonance hybrid of enamine and ylide contributing structures. An alternative proposal for step (c) invokes addition to the carbonyl group but protonation of the intermediate oxyanion by His_{271}. In the final set of reactions, HETTP reacts with enzyme-bound lipoamide (lipoamide-E2p; see Fig. 31.5 above) to form acetyldihydrolipoamide-E2p (steps e–g; His_{128} has been proposed as the proton source for step f).

Leading references

Patel, M. S.; Nemeria, N. S.; Furey, W.; Jordan, F. *J. Biol. Chem.* **289**, 16615–16623 (2014); Kluger, R. and Tittmann, K. *Chem. Rev.* **108**, 1797–1833 (2008); Frank, R. A. W.; Titman, C. M.; Pratap, J. V.; Luisi, B. F.; Perham, R. N. *Science* **306**, 872–876 (2004); Fries, M.; Jung, H.-I.; Perham, R. N. *Biochemistry* **42**, 6996–7002 (2003); Ciszak, E. M.; Korotchkinn, L. G.; Dominiak, P. M.; Sidhu, S.; Patel, M. S. *J. Biol. Chem.* **276**, 21240–21246 (2003).

Chapter 32

Ribonuclease A

32.1 Bovine pancreatic ribonuclease A

Bovine pancreatic ribonuclease A (EC 3.1.27.5; RNase A; RNase I) is a member of a superfamily of pancreatic ribonucleases. It is an endo phosphodiesterase in that it hydrolyzes an internal phosphate linkage of single-stranded RNA into its component nucleotides. It is specific for the cleavage at the 3′-end of a C or U residue (with the 5′ side either a pyrimidine or purine residue) to form a 3′-phosphorylated C or U product (see Fig. 32.1). It is relatively small, consisting of a single polypeptide chain with 124 AA residues, and has historic significance in that it was the first enzyme to be completely sequenced.

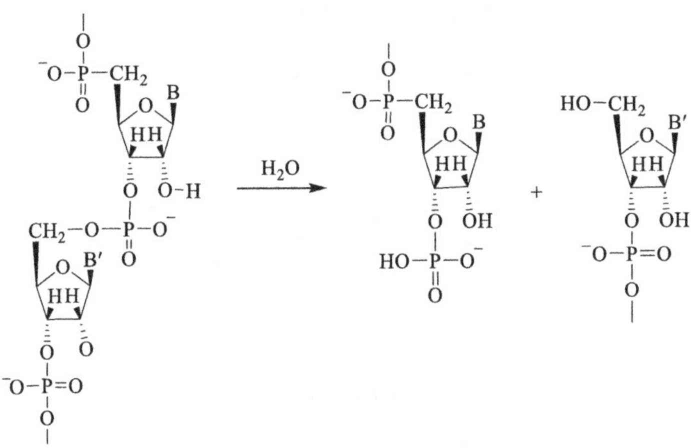

FIGURE 32.1 The overall chemistry catalyzed by RNase A. B=C or U.

32.2 Physiological function

This digestive enzyme plays an important role in RNA metabolism by the cleavage of both native and ingested ribonucleic acid into its component nucleotides. It is ineffective with DNA since the presence of a free 2′-OH group is essential to the hydrolytic mechanism (see below).

Enzyme Active Sites and their Reaction Mechanisms. DOI: https://doi.org/10.1016/B978-0-12-821067-3.00032-5

32.3 Key structural features

The active site lies in a cleft that divides the enzyme into two domains. Subsites (B, R, and P) have been identified that are specific for the (pyrimidine) base, ribose ring, and phosphate groups, respectively. The B_1 site is defined by Val_{43}, Thr_{45}, Phe_{120}, and Ser_{123}. H-bonding of the Thr with a uridine or cytidine and stacking interactions of the base with the Phe provide substantial portions of the stabilizing interactions. Gln_{69}, Asn_{71}, and Glu_{111} likewise stabilize the purines that are preferred for the 5′ side of the scissile bond. Three key catalytic residues, His_{12}, His_{119}, and Lys_{41}, lie in the P_1 site. An image of the active site with uridine 5′-monophosphate bound into the B_1 site is shown in Fig. 32.2.

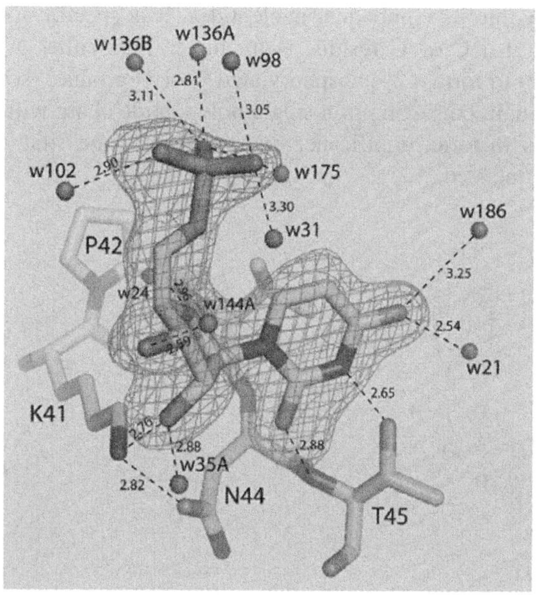

FIGURE 32.2 Uridine 5′ monophosphate bound into the pyrimidine binding site of RNase A. *(Reproduced from Larson et al.* Acta Cryst. ***F66****, 113–120 (2010) with permission of the International Union of Crystallography).*

32.4 Reaction sequence

The hydrolytic cleavage reaction is generally thought to proceed through a cyclic phosphorane intermediate. The sequence is shown in Fig. 32.3. The reaction sequence for the hydrolysis of a phosphate ester may be found in Chapter 28, Phosphatidylinositol-specific phospholipase C.

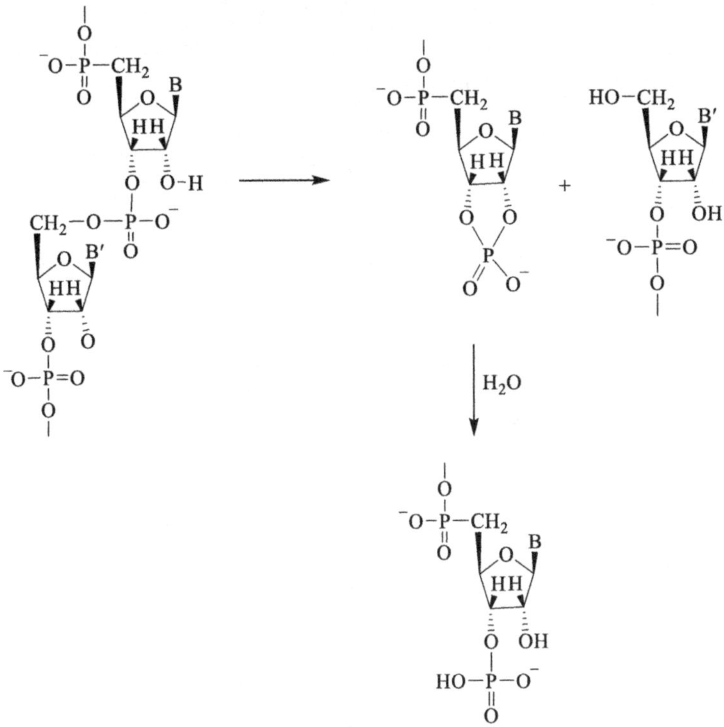

FIGURE 32.3 The reaction sequence for RNase A hydrolytic phosphate cleavage.

32.5 Detailed mechanism including the role of His_{12} and His_{119} at the active site

The commonly accepted mechanism for RNase A is shown in Fig. 32.4. As seen here, the 2′-hydroxyl group plays the role of a catalytic alkoxy group, with the two histidine groups providing general acid–base catalysis. In step (a) His_{12} deprotonates the nucleophile, the 2′-OH group of the 3′-nucleotide, thus initiating an attack on the P=O group. The positively charged Lys_{41} and His_{119} provide electrostatic stabilization of the developing negative charge. The resulting oxyanion is drawn here as an intermediate but has also been suggested as a transition state leading to the product of step (b). When drawn as an intermediate, there has also been discussion of its existence as a mono or dioxy anion (i.e., whether the proton on His_{12} actually resides on the oxygen atom). Cleavage of the 5′ P-O bond in step (b), assisted by a proton transfer from the His_{119}, leads to the formation of the phosphorane. The final hydrolysis in steps (c) and (d), involving a water molecule within the active site, is initiated by the basic His_{119} and catalyzed by the acidic His_{12}. This could proceed as an addition–elimination reaction analogous to steps "a" and "b." It has been suggested, but not generally accepted, that the roles of the Lys_{41} and His_{119} might be reversed. A concerted mechanism has also been proposed.

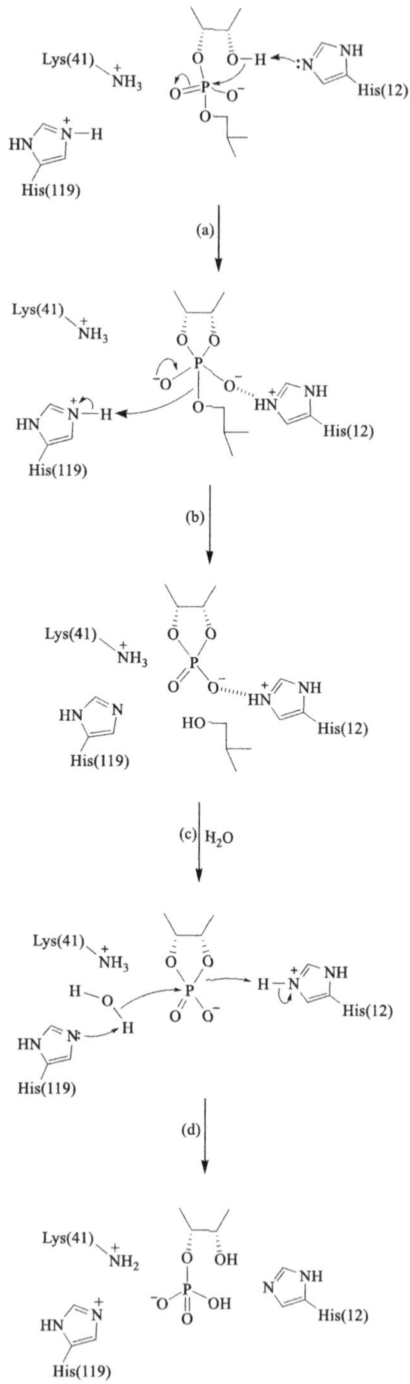

FIGURE 32.4 Detailed mechanism for ribonuclease A and the role of the active site residues. *(Adapted with permission from Elsässer et al.* J. Am. Chem. Soc. ***136****, 927–936 (2014); Copyright 2014, American Chemical Society).*

Leading references

Kasireddy, C.; Ellis, J. M.; Bann, J. G.; Mitchell-Koch, K. R. *Chem. Phys. Lett.* **666**, 58–61 (2016); Harris, M. E.; Piccirilli, J. A.; York, D. M. *Biochim. Biophys. Acta* **1854**, 1801–1808 (2015); Elsässer, B.; Fels, G.; Weare, J. H. *J. Am. Chem. Soc.* **136**, 927–936 (2014); Cuchillo, C. M.; Nogués, M. V.; Raines, R. T. *Biochemistry* **50**, 7835–7841 (2011); Larson, S. B.; Day, J. S.; Nguyen, C.; Cudney, R.; McPherson, A. *Acta Cryst.* **F66**, 113–120 (2010).

Chapter 33

Ribonucleotide reductase

33.1 Ribonucleotide reductase

Ribonucleotide reductase (EC 1.17.4.1; RNR) is a ubiquitous enzyme the primary function of which is to convert ribonucleotides into deoxyribonucleotides. In *Escherichia coli*, an organism that has contributed much of what is known about RNR; the substrates are the ribo di- (and tri) phosphates, that is, ADP, CDP, GDP, and UDP. There are several classes of RNRs. *E. coli*, and most eukaryotes, utilize "class 1a," an aerobic enzyme that contains a pair of ferrous atoms that facilitate the oxidation/reduction chemistry central to the deoxygenation reaction. RNR is one of several groups of enzymes that incorporate iron atoms for oxygen activation. The enzyme consists of two pairs of homodimers—a pair of adjoined subunits labeled α (R_1; 761 AA) interfaced with a pair labeled β (R_2; 375 AA), which associate to form a heterotetramer. Both subunits are required for the (pair of) active sites. Each α component contributes a pair of thiol groups involved in the oxidation–reduction sequence while each β component contributes a pair of oxygen-bridged iron atoms. The enzyme also has an allosteric site in α that controls which ribonucleotide accesses the active site. The overall reaction catalyzed by RNR is shown in Fig. 33.1.

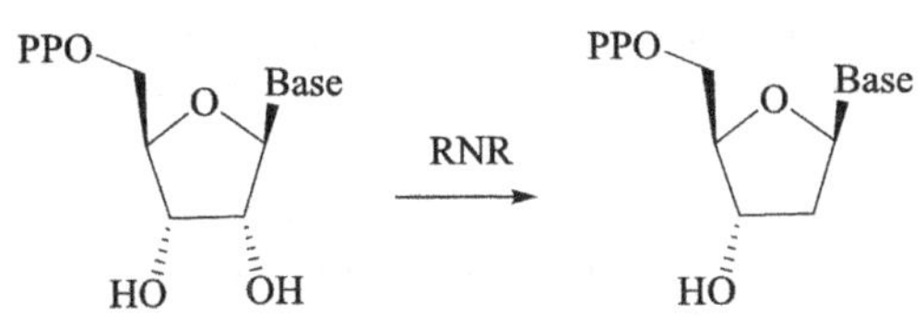

FIGURE 33.1 Overall reaction catalyzed by RNR. *RNR*, Ribonucleotide reductase.

33.2 Physiological function

All but one of the DNA monomers is formed by RNR. The exception is dTMP which is synthesized from dUMP by thymidylate synthase (see Chapter 37, Thymidylate Synthase).

Enzyme Active Sites and their Reaction Mechanisms. DOI: https://doi.org/10.1016/B978-0-12-821067-3.00033-7

33.3 Key structural features

The binuclear, oxygen-bridged ferric iron complex involves six AAs as shown in Fig. 33.2. The reaction is initiated in the β_2 subunit when the

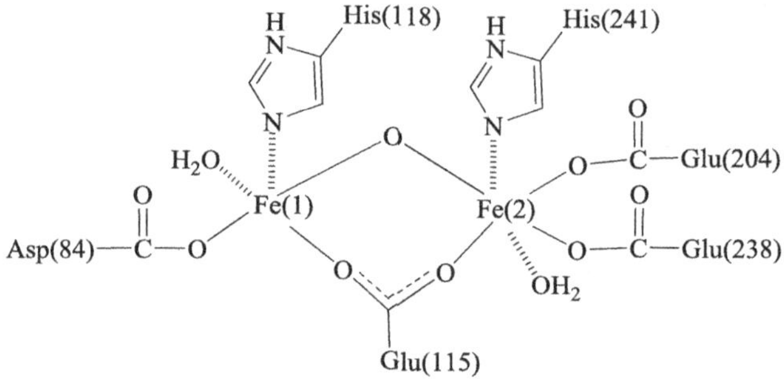

FIGURE 33.2 The binuclear, oxygen-bridged ferric iron complex at the active site of ribonucleotide reductase. *(Adapted with permission of Annual Reviews, Inc., from Cotruvo Jr., J. A.; Stubbe,* J. Annu. Rev. Biochem. ***80****, 733–767 (2011); Copyright 2011; permission conveyed through Copyright Clearance Center, Inc).*

diferrous form (generated by reaction of the diferric form with a reduced flavin), and oxygen, convert a nearby Tyr_{122}-OH to a tyrosyl radical (Tyr_{122}-$O^{\bullet}$) concomitant with the reformation of the oxygen-bridged diferric complex. Though exceedingly short-lived in solution, this radical is sheltered deep in the protein, away from solvent and reactive side chains, and has a half-life of 4 days in that environment.

The tyrosyl radical is not, however, the species that reacts with the ribose ring in the deoxygenation chemistry outlined below in Fig. 33.3. It is too far removed from the substrate. Rather, there is a multistep set of hydrogen atom transfer reactions that ultimately leads to the formation of a Cys(439)-$S^{\bullet}$ in the $\alpha 2$ subunit some 35 Å away. Termed "proton-coupled electron transfer", the chemistry starts with the Tyr_{122} radical and then (consecutively) involves $Tryp_{48}$ and Tyr_{356} in $\beta 2$ and Tyr_{731} and Tyr_{730} in $\alpha 2$. It is the Tyr_{730}-$O^{\bullet}$ that converts Cys_{439}-SH into Cys_{439}-$S^{\bullet}$. Two other pairs of cysteines are also utilized: Cys_{225} and Cys_{462} on $\alpha 2$ are oxidized to form a disulfide bond during the ribose reduction chemistry. These are then returned to their dithio form through the involvement of a second pair, Cys_{734} and Cys_{759} on $\alpha 2$, which shuttle electrons to the active site for this reduction. The ultimate source of these electrons is NADP(H) via an enzyme, thioredoxin reductase. Glu_{441} acts as a base in one suggested mechanism.

33.4 Reaction sequence

The reduction mechanism is outlined in broad terms in Fig. 33.3. Of particular note is the dehydration step. It has intentionally been left somewhat vague because there are several options for the detailed sequence. Two of these are shown below in Figs. 33.3 and 33.4. They differ in the way by

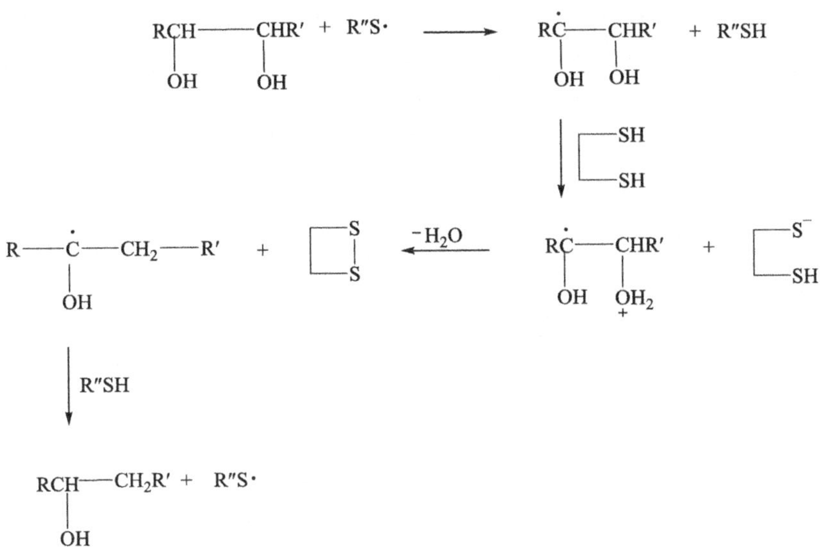

FIGURE 33.3 General mechanism for the ribonucleotide reductase-catalyzed reduction of a vicinal diol to a mono alcohol.

which Cys_{225} and Cys_{462} transfer electrons and hydrogen atoms to the ribose ring and in the role of Glu_{441}.

33.5 Detailed mechanisms and the role of the active site residues

This is shown in Fig. 33.4. The reaction is initiated in step (a) with hydrogen abstraction from the C3′ carbon by the Cys_{439} thiyl radical. The 2′ hydroxyl group is protonated in step (b) by Cys_{225}. The subsequent loss of water in step (c), concomitant with deprotonation of the C3′ OH group by the glutamate anion, generates an alpha carbocation ketyl species which can be drawn in an alternative resonance structure as an alpha keto radical. In step (d) a hydrogen atom is transferred from the Cys(462) thiol group with the formation of a disulfide radical anion. An electron is transferred to the keto group in step (e) and the ketyl species is converted to product in step (f).

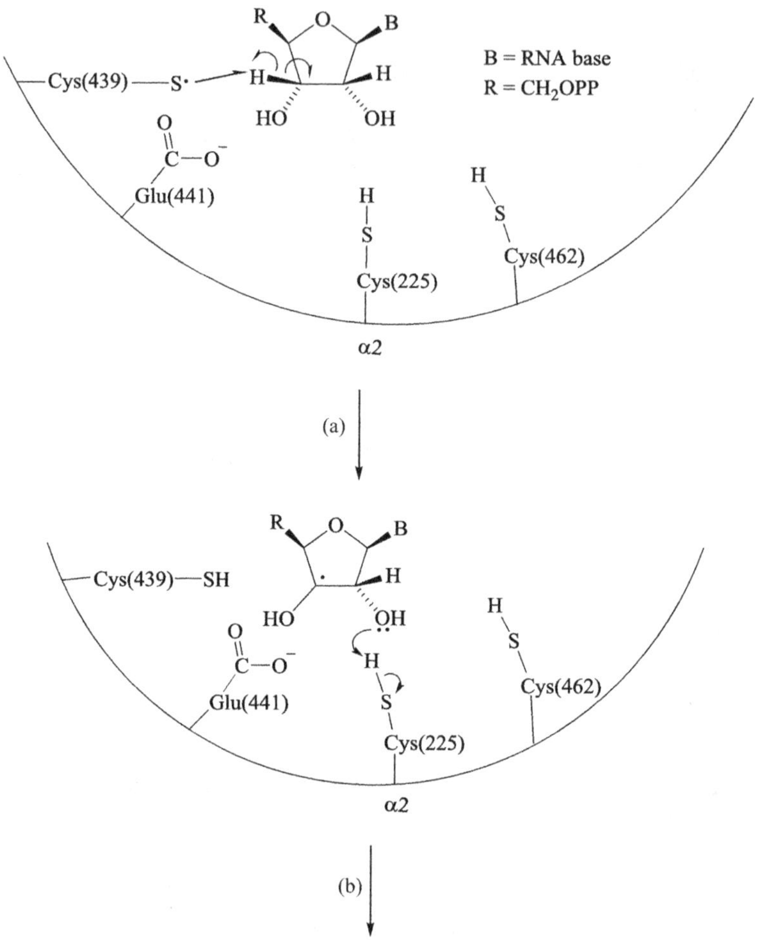

FIGURE 33.4 Detailed mechanism for the ribonucleotide reductase-catalyzed conversion of a ribose ring to a deoxyribose ring showing the involvement of key amino acids at the active site. *(Adapted with permission of Annual Reviews, Inc.; Copyright 2011. See Fig. 33.2).*

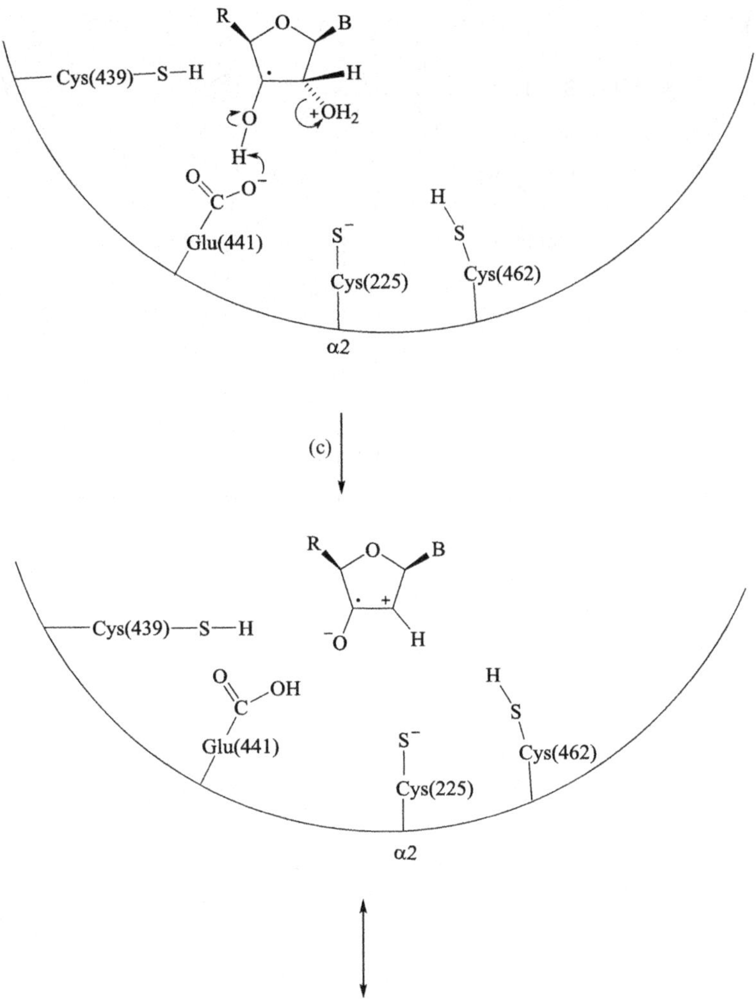

FIGURE 33.4 (Continued).

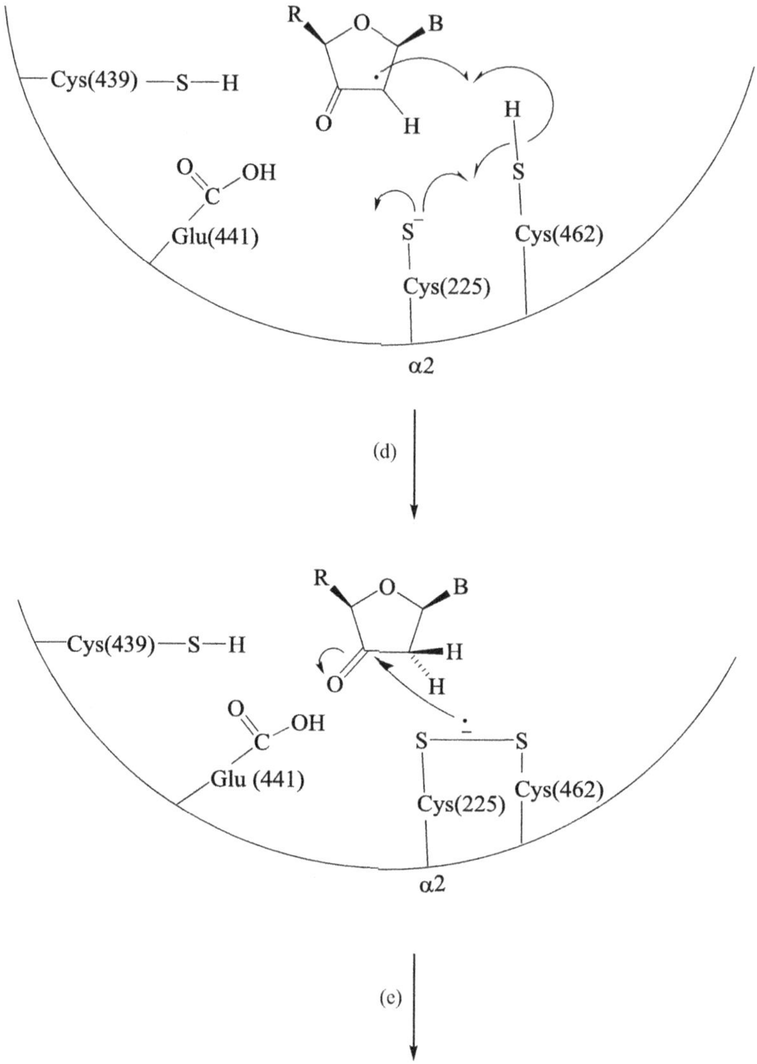

FIGURE 33.4 (Continued).

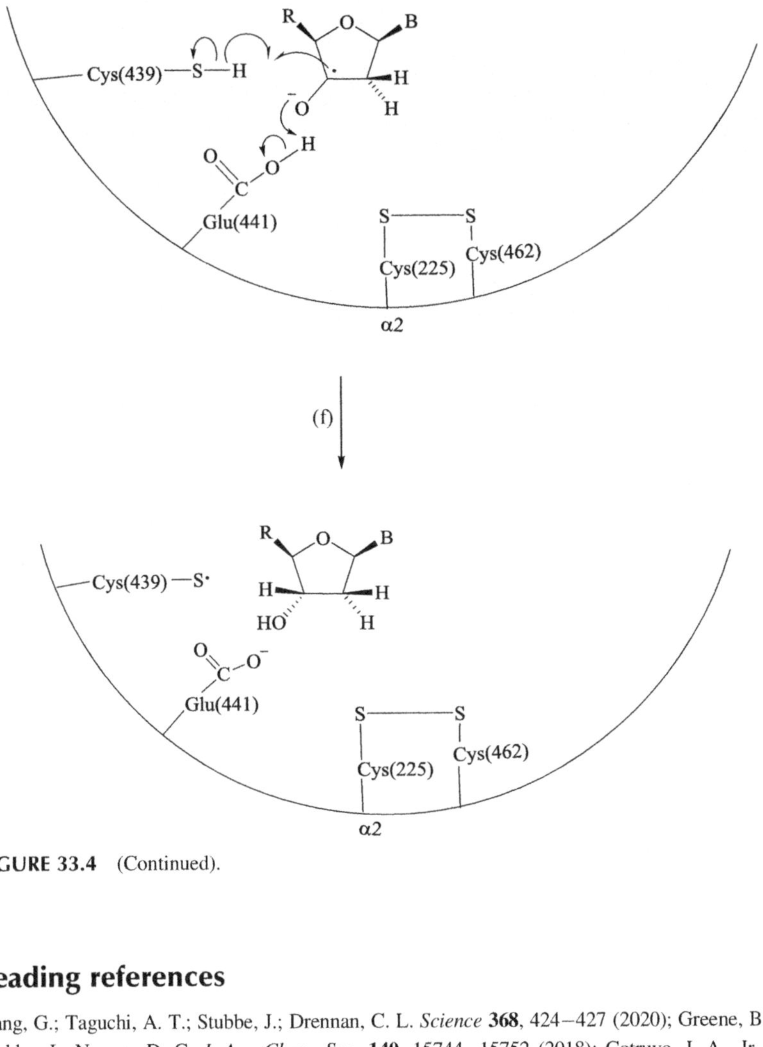

FIGURE 33.4 (Continued).

Leading references

Kang, G.; Taguchi, A. T.; Stubbe, J.; Drennan, C. L. *Science* **368**, 424–427 (2020); Greene, B. L.; Stubbe, J.; Nocera, D. G. *J. Am. Chem. Soc.* **140**, 15744–15752 (2018); Cotruvo, J. A., Jr. and Stubbe, J. *Annu. Rev. Biochem.* **80**, 733–767 (2011); Holmgren, A. and Sengupta, R. *Free Radic. Biol. Med.* **49**, 1617–1628 (2010); Kolberg, M.; Strand, K. R.; Graff, P.; Andersson, K. K. *Biochim. Biophys. Acta* **1699**, 1–34 (2004); Jordan, A. and Reichard, P. *Annu. Rev. Biochem.* **67**, 71–98 (1998).

Chapter 34

Serine racemase

34.1 Serine racemase

Serine racemase (EC 5.1.1.18; SR; SerR) is an enzyme which has, as its primary function, the conversion of L-serine (S isomer) to D-serine (R isomer). It also catalyzes the dehydration of both serine enantiomers to pyruvate and ammonia. Enzyme activity is enhanced by the presence of a divalent metal cation (e.g., Ca^{2+}, Mg^{2+}, and Mn^{2+}) and Mg · ATP, though neither of these is present at the active site. Pyridoxal 5′-phosphate (PLP) is a requisite cofactor (see Fig. 34.1). The overall reaction catalyzed by SerR is shown in Fig. 34.2.

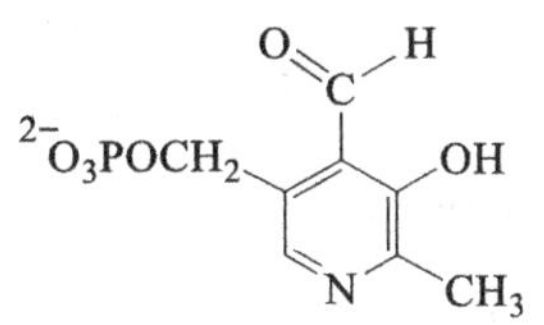

FIGURE 34.1 Structure of PLP. *PLP*, Pyridoxal 5′-phosphate.

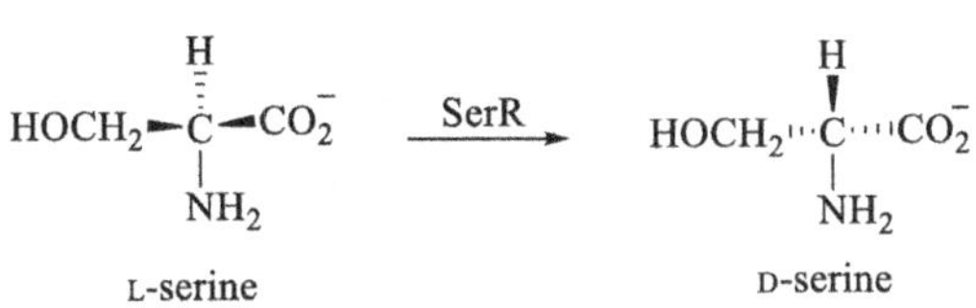

FIGURE 34.2 The overall reaction catalyzed by SerR.

34.2 Physiological function

The highest level of D-serine is found in the brain, though it is present, for example, in mammalian skin. D-serine is critical for the activation of *N*-methyl-D-aspartate receptors (NMDARs) in the nervous system. The

Enzyme Active Sites and their Reaction Mechanisms. DOI: https://doi.org/10.1016/B978-0-12-821067-3.00034-9

NMDARs have two binding sites—glutamate and glycine/D-serine. The activation of these receptors has been found to be important for neurotransmission, synaptic plasticity, learning, and memory, but excessive activation is associated with Alzheimer, Parkinson, and Huntington diseases, as well as with amyotrophic lateral sclerosis and stroke. Conversely, a low level of activation is associated with schizophrenia. The competition between racemization to, and the dehydration of, D-serine by SerR provides the requisite balance of D-serine concentration.

34.3 Key structural features

SerR is a cystolic, globular, dimeric protein. PLP-dependent enzymes have been grouped into five classes based on their secondary structures—SerR is assigned to the fold-type II grouping. Two domains, "large" and "small," contribute to the active site, which is in an "open" conformation absent of the substrate and "closed" when the site is occupied. The conformational changes caused by the ligand at the active site occur primarily within the (relatively flexible) small domain. Neither of the mechanisms by which the allosteric activators, Mg · ATP and an additional divalent metal cation, operate has been delineated. The (modeled) ATP binding site utilizes amino acids from both monomeric units that are typical for an ATP binding site. A partial list includes Tyr_{121} (which stacks with the adenosine ring) and Gln_{50}, Ly_{51}, Arg_{277}, and Lys_{279}, which coordinate with the Mg^{2+}, γ-phosphate, adenosine ring, and 3′-OH of the ribose ring, respectively. It is remote from the active site but appears to be connected with it through a network of hydrogen bonds. The divalent metal binding site is closer to (within one AA of) the active site where the metal is coordinated with Glu_{210}, Ala_{214}, Asp_{216}, and three molecules of water. The key amino acids at the active site for human SerR are shown in Fig. 34.3. Note that PLP is

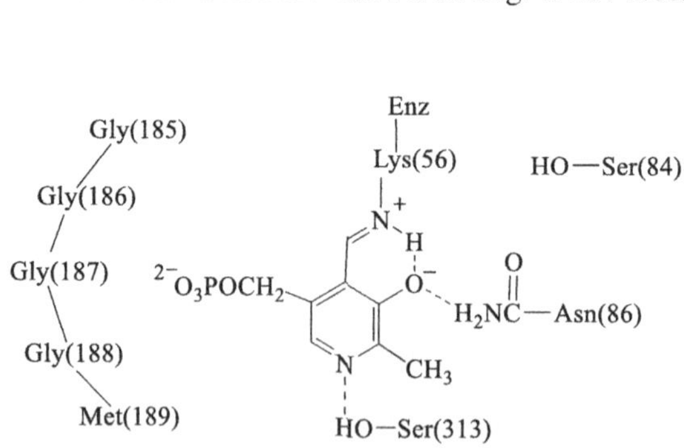

FIGURE 34.3 Binding site for PLP in human SerR. *PLP*, Pyridoxal 5′-phosphate.

bound to the enzyme through Lys_{56}. A "glycine cluster" coordinates to the phosphate group.

34.4 Reaction sequence

The racemization (Fig. 34.4) involves, first, the exchange of L-serine for the amine component of an enzyme-linked Schiff-base (steps a and b). The "alpha" carbon of the new Schiff-base is then deprotonated by a base to form an anion (step c). Reprotonation from the opposite face (step d) forms the enantiomeric Schiff-base, which then reexchanges with the original amine to release D-serine (step e). The biggest hurdle for this chemistry is the deprotonation of a normally relatively nonacidic C—H bond (step c).

34.5 Detailed mechanism and the role of active site residues

The mechanism shown below (Fig. 34.5) is similar to others that utilize the PLP cofactor. Not shown is the initial state of the active site, with PLP bound to the enzyme through a Schiff-base involving Lys_{56} (see Fig. 34.3). This aldimine exchanges with L-serine (as shown in Fig. 34.4) to form a new Schiff-base, the starting point for Fig. 34.5. Note that such an exchange reaction is likely to be base-catalyzed but no specific base at the active site has been assigned this role. In the highly analogous mechanism for serine dehydratase, it is the PLP phosphate anion that is proposed to initiate the deprotonation.

In step a, the amino group of Lys_{56} initiates the deprotonation of the L-serine α-carbon. The Lys_{56} cation is stabilized by the carbonyl group of Pro_{153} (not shown). Though such deprotonation is typical of the chemistry of PLP-bound substrates, the novel feature here is that there is no evidence that the pyridine nitrogen is protonated, as is typically the case. This is attributed to the strong hydrogen bond with the nearby Ser_{313}. Thus stabilization of the carbanion by delocalization of the negative charge onto a pyridinium species to form a quinoidal contributor is not available here. Alternatively, the carbanion stabilization has been attributed to solvation by water and active-site AAs, and by electrostatic interactions with the enzyme.

In step b, the carbanion is protonated on the opposite face by Ser_{84}. As with the Lys_{56} cation, the Ser_{84}-O^- anion is stabilized primarily by solvation by water. Step c proceeds through the detailed transaldimation chemistry of Fig. 34.4 to reform the enzyme-bound PLP and the release of D-serine.

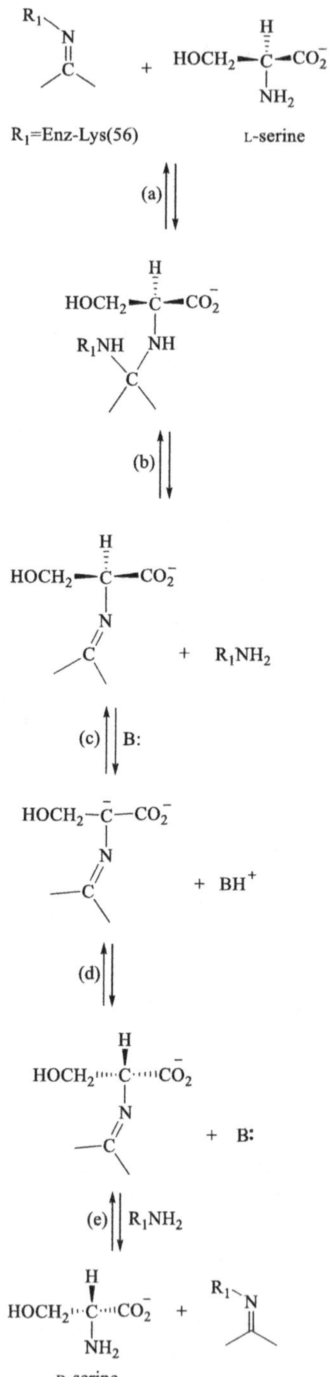

FIGURE 34.4 Chemical reaction sequence for the interconversion of serine enantiomers through the intermediacy of an aldimine (Schiff-base).

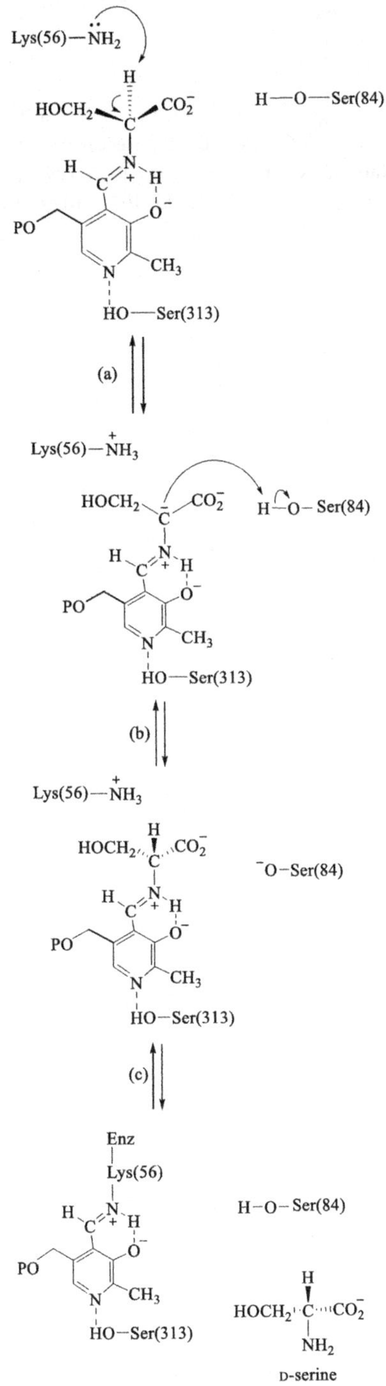

FIGURE 34.5 Detailed mechanism for, and the role of active site residues in, the racemization of L-serine by SerR. *(Adapted with permission from Nitoker, N.; Major, D. T.* Biochemistry ***54****, 516–527 (2015); Copyright 2015 American Chemical Society).*

Leading references

Liang, J.; Han, Q.; Tan, Y.; Ding, H.; Li, J. *Front. Mol. Biosci.* **6**, 1–21 (2019); Nelson, D. L.; Applegate, G. A.; Beio, M. L.; Graham, D. L.; Berkowitz, D. B. *J. Biol. Chem.* **292**, 13986–14002 (2017); Nitoker, N. and Major, D. T. *Biochemistry* **54**, 516–527 (2015); Jirásková-Vaničková, J., et al. *Curr. Drug Targets* **12**, 1037–1056 (2011).

Chapter 35

Soluble quinoprotein glucose dehydrogenase

35.1 Soluble quinoprotein glucose dehydrogenase

Soluble quinoprotein glucose dehydrogenase [EC 1.1.99.35 (earlier: 1.1.19.17; sGDH; s-GDH)] is a bacterial enzyme that oxidizes a wide range of aldoses to lactones. The enzyme has attracted considerable attention, not only because of its use of a novel cofactor (see below) but also as a means to monitor blood glucose levels. An example of the overall reaction initiated by sGDH is shown in Fig. 35.1, wherein β-D-glucose (see Fig. 35.2) is oxidized to D-glucono-δ-lactone (which is hydrolyzed to gluconic acid).

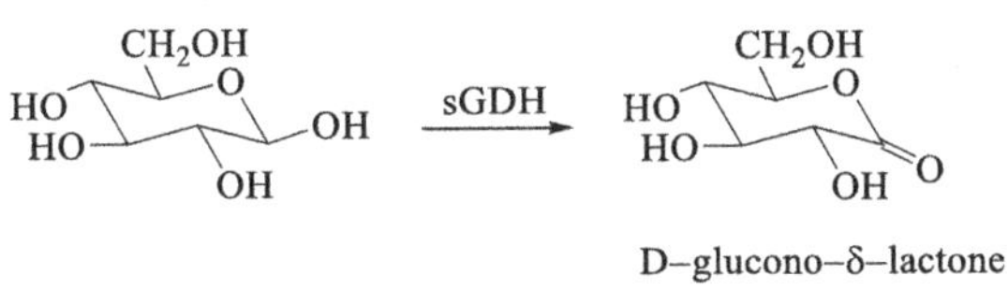

FIGURE 35.1 The overall reaction catalyzed by sGDH.

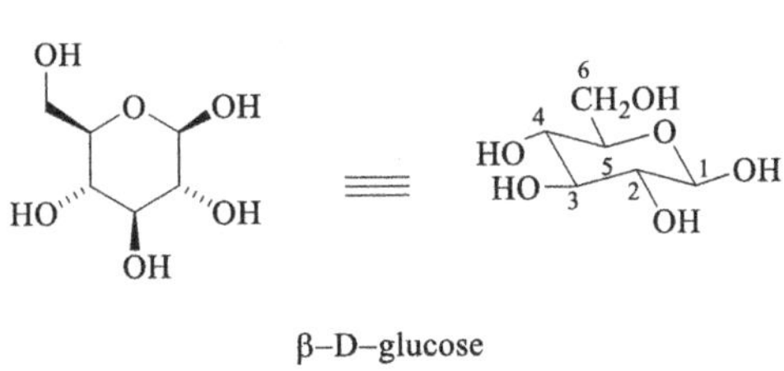

FIGURE 35.2 Structural representations of β-D-glucose.

sGDH utilizes pyrroloquinoline quinone (methotaxin, PQQ) as a cofactor for this oxidation. PQQ, and several related quinoidal cofactors, enable a group of quinoproteins that constitute a third class of oxidoreductases (in addition to those that employ NAD and flavins). During the oxidation PQQ is reduced to a dihydroquinone, pyrroloquinoline quinol ($PQQH_2$) (see Fig. 35.3). Calcium ions are also required for enzymatic activity.

Enzyme Active Sites and their Reaction Mechanisms. DOI: https://doi.org/10.1016/B978-0-12-821067-3.00035-0

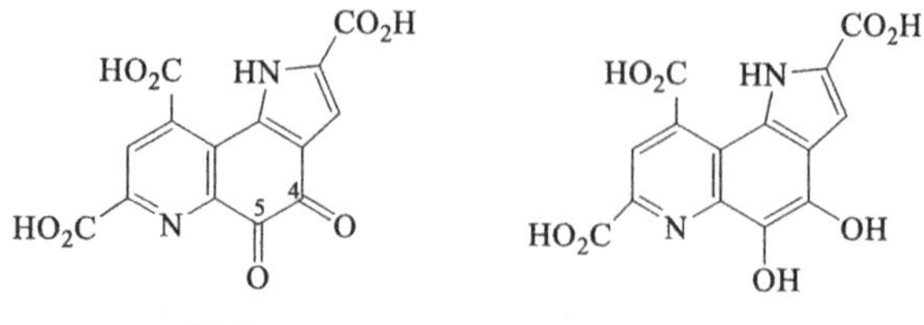

FIGURE 35.3 The structures of PQQ and $PQQH_2$. *PQQ*, pyrroloquinoline quinone; *$PQQH_2$*, pyrroloquinoline quinol.

35.2 Physiological function

The primary source of sGDH has been *Acinetobacter calcoaceticus* where it is found in the periplasmic space (a membrane-bound, but structurally different, GDH (mGDH) has also been isolated from this bacterium). Glucose is oxidized as a key step in bacterial respiration; the biological entity that oxidatively regenerates PQQ has not yet been identified. sGDH chemistry is insensitive to oxygen, a major motivation for the use of this enzyme in glucose sensor devices.

35.3 Key structural features

sGDH is a dimer consisting of a pair of identical subunits. These are adjoined with the aid of a pair of calcium ions. In addition, when activated, each of these contains a third calcium ion and a molecule of PQQ within the active site. Four water molecules, Gly_{247}, and Pro_{248} are found bound to the calcium ion in a crystal lacking PQQ. An X-ray structure with the presence of PQQ shows the calcium ion to now be hepta-coordinate, binding to the O5, N6, and O7A of PQQ, Gly_{247}, and Pro_{248}, and two water molecules (see Fig. 35.4). The X-ray structure

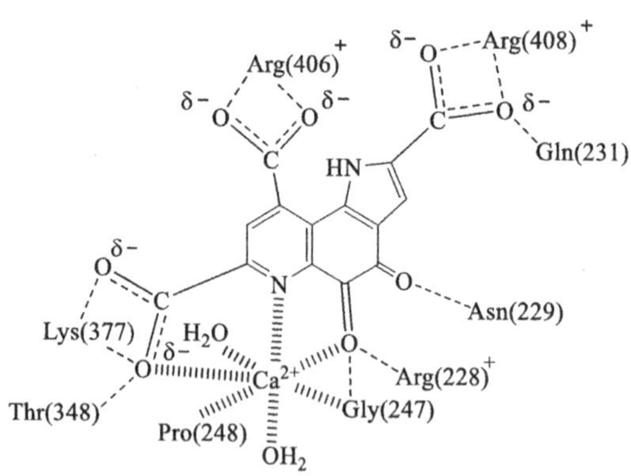

FIGURE 35.4 Binding of Ca^{2+} and PQQ to sGDH at the active site. *sGDH*, Soluble quinoprotein glucose dehydrogenase; *PQQ*, pyrroloquinoline quinone. *(Adapted with permission from Reddy, S. Y.; Bruice, T. C. J.* Am. Chem. Soc. ***126****, 2431–2438 (2004); Copyright 2004 American Chemical Society).*

indicates that the Pro_{248} detaches from the calcium atom after reduction forms an alkoxide anion at C5 (see mechanism below), but reassociates with it after protonation of the O5 anion. At that point, the calcium atom is no longer coordinated with O5 but has added a third water molecule and is again hepta-coordinate.

The PQQ C2, C7, and C9 carboxylate groups form salt bridges with Arg_{408}, Lys_{377}, and Arg_{406}, respectively (Fig. 35.4). The quinone sits on a hydrophobic "shelf" made up of Gln_{231}, Gln_{246}, Ala_{350}, and Leu_{376} (not shown).

When bound at the active site, the β conformer of D-glucose sits above the PQQ with its axial hydrogen at C1 pointing toward the C5 carbonyl group of the cofactor (it is this feature that confers the high level of stereospecificity exhibited by this enzyme). The positioning of the glucose is guided by hydrophobic interactions with the PQQ surface and with Leu_{169}, Tyr_{343}, and $Tryp_{346}$. H-bonding by glucose with the protein is limited to the O1 and O2 hydroxyl groups. The former interacts with His_{144}, Gln_{168}, and Arg_{228}, all available only when the —OH group is equatorial (see comment above about the enzyme's stereospecificity). The O2 hydroxyl group forms H-bonds with Gln_{76} and Asp_{143}.

35.4 Reaction sequence

The reaction depicted in Fig. 35.1 involves a reduction of PQQ at C5 concomitant with the oxidation at the O2 position of the glucose. The reduced PQQ then enolizes to form the enediol, PQQH2. This sequence is shown in Fig. 35.5.

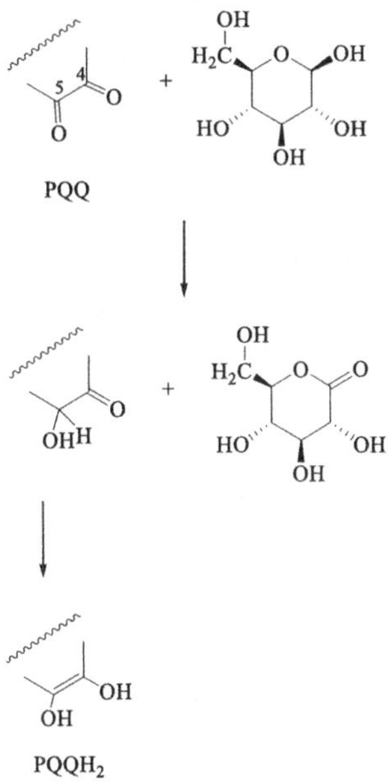

FIGURE 35.5 The proposed reaction sequence for the formation of D-glucono-δ-lactone by sGDH. *sGDH*, Soluble quinoprotein glucose dehydrogenase.

35.5 Detailed mechanism and the role of active-site residues

The basic elements of the mechanism are now well understood. This has been achieved through X-ray structures and subsequent molecular dynamics simulations. The key oxidation/reduction step involves a hydride transfer from the glucose to C5 of the PQQ. A detailed sequence is shown in Fig. 35.6.

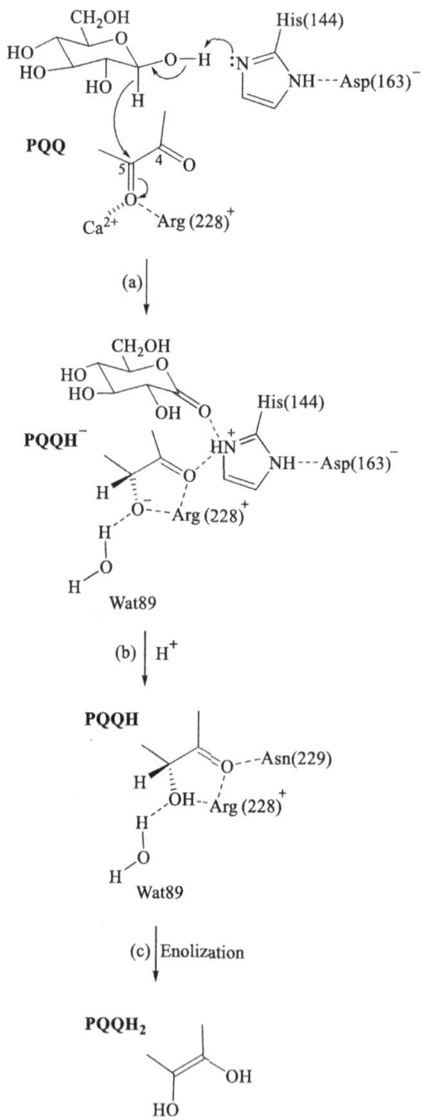

FIGURE 35.6 Detailed mechanism for soluble quinoprotein glucose dehydrogenase showing the involvement of key amino acids at the active site.

In step (a) His_{144} (assisted by Asp_{163}) acts as a base to deprotonate the glucose C1 hydroxyl group. This leads to a transfer of the C1 hydrogen, as a hydride anion, to the C5 carbonyl group of PQQ. The electrophilicity of the carbonyl group is enhanced by its coordination to the requisite calcium cation and its interaction with Arg_{228}. The product C5 oxyanion is stabilized by the Arg_{228} and by one of the two active-site water molecules, termed Wat_{89}. The other is Wat_{55}, which actually moves away from the PQQ upon its conversion to $PQQH^-$. Though not shown, the calcium ion remains coordinated to the C5 oxygen on $PQQH^-$ (as noted above, the Pro_{248} has detached at this stage).

Step (b) involves protonation of the oxyanion to form the alcohol (PQQH; referred to in the literature as the "fluorescent intermediate"). The proton-donating agent is yet to be determined. Though the now-protonated His_{144} has been proposed, simulations indicate it is too far from the oxygen anion to do this directly, and either Wat_{89} or a hydrogen-bond network involving the protonated His_{144} are potential candidates. Again not shown is the calcium ion coordination that continues throughout this process. Simulations suggest that a third water molecule replaces the calcium interaction with O5. Pro_{248} is once again coordinated to the calcium at this stage.

Step (c) is the enolization step that completes the transformation of PQQ to the fully aromatized $PQQH_2$ (see Fig. 35.3). The details here are also murky. His_{144} is again deemed to be too far from the H_5 hydrogen to act as a base. Wat_{89} has been suggested as an alternative base, with aromatization and the O5/O4 interactions with Arg_{228} and Asn_{229} driving this tautomerization. If His_{244} is still protonated at this point, its proximity to the O4 of PQQH would make it a possible candidate as the proton donor to this oxygen during enolization.

Leading references

Duine, J. A.; Strampraad, M. J. F.; Hagen, W. R.; de Vries, S. *FEBS J.* **283**, 3604–3612 (2016); Yoo, E.-H. and Lee, S.-Y. *Sensors* **10**, 4558–4576 (2010); Reddy, S. Y. and Bruice, T. C. *Protein Sci.* **13**, 1965–1978 (2004); Reddy, S. Y. and Bruice, T. C. *J. Am. Chem. Soc.* **126**, 2431–2438 (2004); Oubrie, A. and Dijkstra, B. W. *Protein Sci.* **9**, 1265–1273 (2000); Oubrie, A.; Rozeboom, H. J.; Kalk, K. H.; Olsthoorn, A. J. J.; Duine, J. A.; Dijkstra, B. W. *EMBO J.* **18**, 5187–5194 (1999); Oubrie, A.; Rozeboom, H. J.; Dijkstra, B. W. *Proc. Natl. Acad. Sci. U.S.A.* **96**, 11787–11791 (1999).

Chapter 36

Tetrachloroethene reductive dehalogenase—PceA

36.1 PceA

PceA (PCE-RDase; EC 1.97.1.8) is one of a large family of cobalt-dependent (reductive) dehalogenases (RDases) that are involved in the utilization of halogenated, aliphatic or aromatic, hydrocarbons as a source of energy for growth. RDases are found in anaerobic "halorespiring bacteria" (also organohalide-respiring bacteria) and have the common capability of transforming a substrate carbon-halogen bond into a C-H bond concomitant with loss of the halogen atom. Specifically, PceA reduces tetrachloroethylene (perchloroethylene) to trichloroethylene (TCE), and then further to *cis*-dichloroethylene (*cis*-DCE; Z-DCE) (see Fig. 36.1).

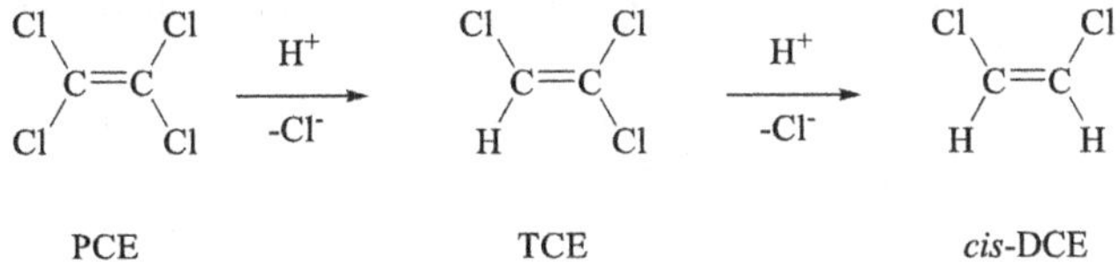

FIGURE 36.1 The chemical transformations catalyzed by PceA.

36.2 Physiological function

There are a number of bacteria that utilize "halorespiration," wherein the dehalogenation reaction serves as the terminal step. Though halogenated organic compounds are naturally occurring in the environment, these enzymes have attracted considerable interest because of their potential use in the bioremediation of areas heavily contaminated by man-made halogenated solvents and other halogen-bearing industrial byproducts. The best-characterized (see below) PceA has been isolated from *Sulfurospirillum multivorans*, where it is found associated with the cytoplasmic membrane.

36.3 Key structural features

Two features of the PceA active site characterize the vast majority of the RDase family. One is the presence of a corrinoid core (here a vitamin B-12-like

Enzyme Active Sites and their Reaction Mechanisms. DOI: https://doi.org/10.1016/B978-0-12-821067-3.00036-2

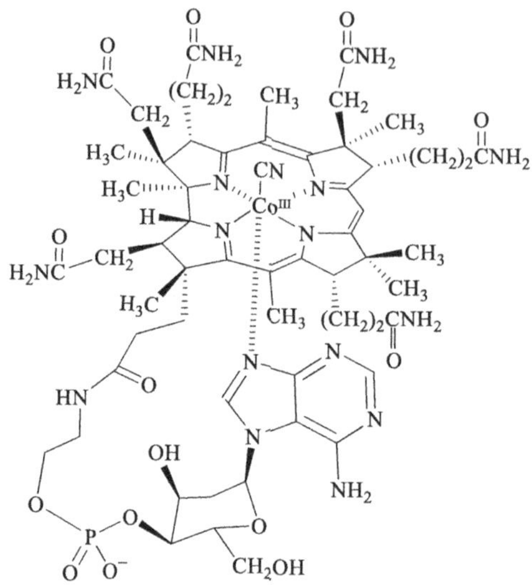

FIGURE 36.2 The structure of norpseudo-B_{12}.

"norpseudo-B_{12}"; see Fig. 36.2; see also Chapter 24, Methylmalonyl Coenzyme A Mutase). The other is the involvement of two iron–sulfur [4Fe-4S] clusters. In addition to the RDase family, one finds vitamin B-12 in adenosylcobolamin-dependent isomerases and methylcobolamin-dependent methyltransferases.

There are several published crystal structures of PceA from *S. multivorans* including: (1) the enzyme without substrate, (2) the enzyme with bound TCE, and (3) the enzyme with a bound *cis*-DCE analog, *cis*-dibromoethene. The enzyme is a dimer containing two independent active sites. As isolated without substrate, the Co exists in the +2 oxidation state (Co^{II}), with the β (upper) axial ligand tentatively identified as a water molecule, and the lower ligand in a "base-off" configuration. A 12 Å hydrophobic channel leads the substrate to the active site. A group of AA's constrain the entrance to this channel; termed a "letter box" these consist of Thr_{39}, Phe_{44}, Phe_{57}, Leu_{186}, and Glu_{189}. Sitting just above the cofactor is a "perimeter fence" of Arg305, Asn272, Trp376, and Phe38, which restricts access to the Co to a water molecule. These are shown in Fig. 36.3, together with Tyr_{246}, which is invoked in the mechanism (see below), and the two iron–sulfur clusters. One cluster is relatively close to the core (labeled "proximal") and the other more distant ("distal").

Upon entry into the active site, the TCE substrate occupies a position above the Co, with one of the two geminal chlorine atoms pointed toward the metal atom. Two nearby AA's that likely play a role in the dehalogenation chemistry are Arg_{305} and Tyr_{246} (see Fig. 36.4). Both are heavily conserved in RDases.

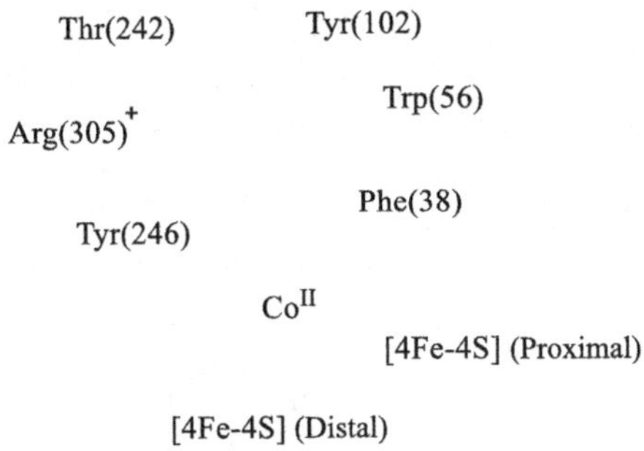

FIGURE 36.3 The key features of the active site of PceA from *S. multivorans*.

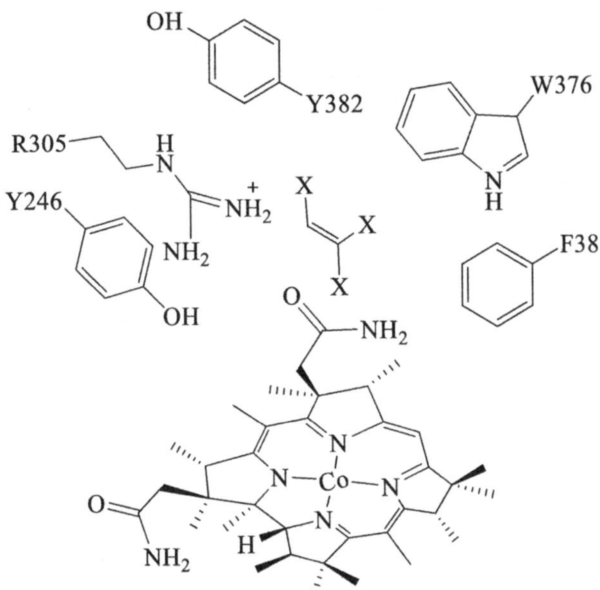

FIGURE 36.4 The active site of PceA *S. multivorans* with bound TCE substrate. *TCE*, Trichloroethylene. *(Adapted by a Creative Commons License from Johannissen, L. O. et al.* Phys. Chem. Chem. Phys. ***19****, 6090–6094 (2017); courtesy of the Royal Society of Chemistry).*

36.4 Reaction sequence

As indicated by the name of this class of enzymes, the "reductive dehalogenation" is initiated by electron transfer from the (superoxidized) cobalt(I) ion to the halogenated alkene. There are several proposals for the details of the following chemistry (see below) but ultimately (for TCE) one of the two geminal halogen atoms is lost and DCE is formed, stereospecifically as the cis (Z) isomer (see Fig. 36.5).

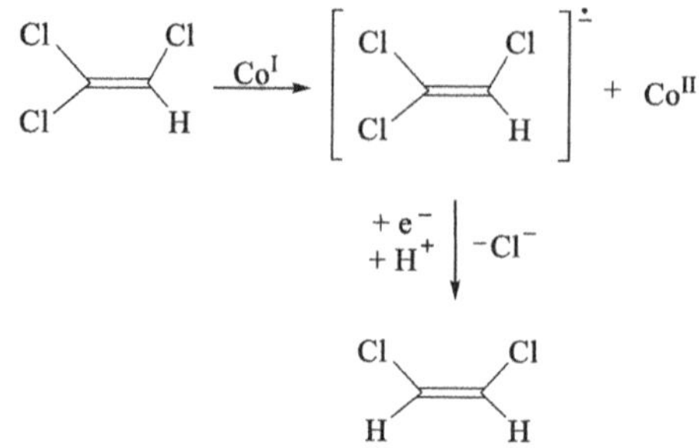

FIGURE 36.5 PceA dechlorination of TCE to *cis*-DCE. *TCE*, Trichloroethylene; *cis*-DCE, *cis*-dichloroethylene.

36.5 Detailed mechanism and the role of active-site residues

There is broad consensus that the cofactor, Co(II), is oxidized to Co(I) by the proximal Fe-S cluster. Then, as noted above, electron transfer to TCE leads to the trichloroalkene radical-anion [step a in Fig. 36.6]. The nearby

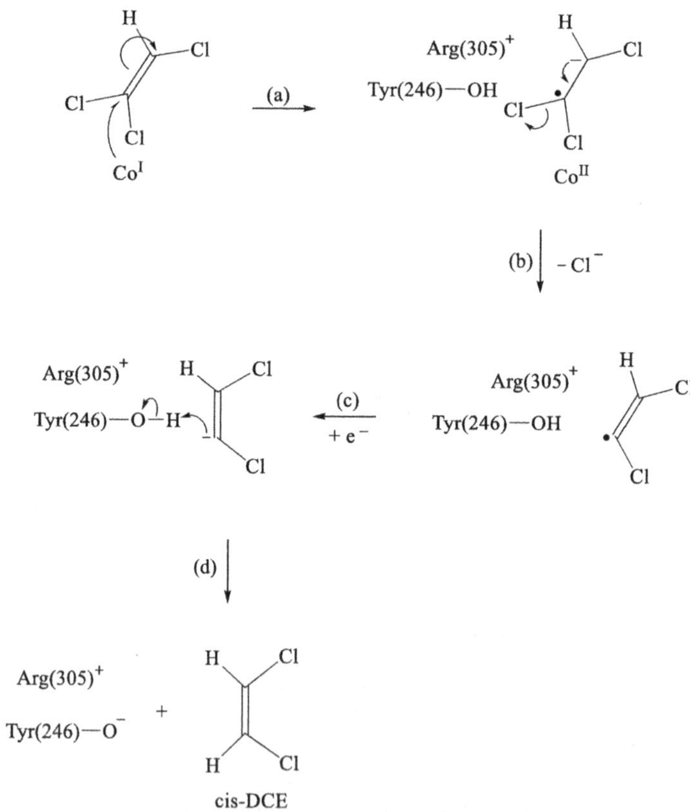

FIGURE 36.6 Mechanism and the role of active-site residues in PceA.

Arg_{305} helps stabilize the developing negative charge. Several mechanisms have been proposed for the subsequent formation of the dichloroalkene. Several invoke the intermediate formation of a C-Co bond. Though this is a viable possibility for the analogous chemistry observed with NpRdhA, such bond formation has been deemed unlikely for PceA because of steric considerations. Alternatively, the radical-anion would be expected to release a chloride anion—either concomitant with the electron transfer or in a step-wise manner [the latter is shown in Fig. 36.6 as step b]. Though one could also portray this cleavage chemistry as homolytic, with the formation of a vinyl carbanion and a chlorine atom, a characteristic rearrangement seen with a substrate bearing an adjacent cyclopropane group gives evidence for an intermediate vinyl radical. A second electron transfer (perhaps involving the distal Fe-S cluster) converts the vinyl radical to a vinyl carbanion (step c). This is then protonated by Tyr_{246} to form *cis*-DCE (step d; computations suggest that the electron transfer and protonation might also be concerted). The nearby Arg_{305} is expected to increase the acidity of the phenolic hydroxyl group. Steric factors have again been invoked to explain the observed stereospecificity of this reaction.

Leading references

Schubert, T.; von Reuß, S. H.; Kunze, C.; Paetz, C.; Kruse, S.; Brand-Schön, P.; Nelly, A. M.; Nüske, J.; Diekert, G. *Microb. Biotechnol.* **12**, 346–359 (2019); Wang, S.; Qiu, L.; Liu, X.; Xu, G.; Siegert, M.; Lu, Q.; Juneau, P.; Yu, L.; Liang, D.; He, Z.; Qiu, R. *Biotechnol. Adv.* **36**, 1194–1206 (2018); Kunze, C.; Diekert, G.; Schubert, T. *FEBS J.* **284**, 3520–3535 (2017); Kunze, C.; Bommer, M.; Hagen, W. R.; Uksa, M.; Dobbek, H.; Schubert, T.; Diekert, G. *Nat. Commun.* **8**, 15858–15869 (2017); Johannissen, L. O.; Leys, D.; Hay, S. *Phys. Chem. Chem. Phys.* **19**, 6090–6094 (2017); Fincker, M. and Spormann, A. M. *Annu. Rev. Biochem.* **86**, 357–386 (2017); Liao, R.-Z.; Chen, S.-L.; Siegbahn, P. E. M. *Chemistry* **22**, 12391–12399 (2016); Adrian L.; and Löffler F. E., Eds; Organohalide-Respiring Bacteria, Springer Verlag: Berlin, Heidelberg, 2016 (); Bommer, M.; Kunze, C.; Fesseler, J.; Schubert, T.; Diekert, G.; Dobbek, H. *Science* **346**, 455–458 (2014).

Chapter 37

Thymidylate synthase

37.1 Thymidylate synthase

Thymidylate synthase (EC 2.1.1.45; TS; TSase) is a critical enzyme for the synthesis of DNA—it converts dUMP (2′-deoxyuridine-5′-monophosphate) into dTMP (2′-deoxythymidine-5′-monophosphate). The source of the C-5 methyl group is a requisite cofactor, (6*R*)-N^5, N^{10}-methylene-5,6,7,8-tetrahydrofolate (MTHF; CH_2H_4fol; 5,10-methylenetetrahydrofolate; see the chapter on deoxyribodipyrimidine photolyase for the structure of MTHF). During the course of this multistep reaction, MTHF is converted into 7,8-dihydrofolate (DHF; H_2folate; see Fig. 37.1). DHF is, in turn, reduced by DHF reductase/NADPH and converted

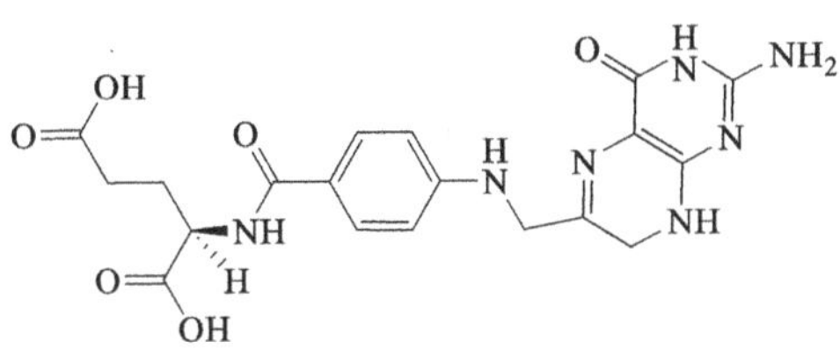

FIGURE 37.1 Structure of DHF. *DHF*, 7,8-Dihydrofolate.

back to MTHF by serine *trans*-hydroxymethylase. Several other enzymes catalyze the transfer of a one-carbon unit to the 5-position of pyrimidines—dCMP and dUMP hydroxymethylases and uracil and cytosine methyl transferases (see elsewhere in this book for m^5C-cytosine methyltransferase). Note the para-aminobenzoic acid and glutamate portions of MTHF, and the CH_2 ("methylene") bridge between N_5 and N_{10} that ultimately is transferred to the uracil. The overall reaction is shown in Fig. 37.2.

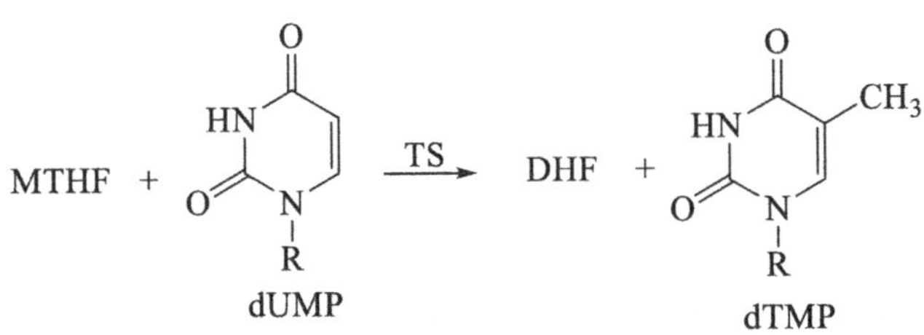

FIGURE 37.2 The overall reaction catalyzed by TS. *TS*, Thymidylate synthase.

Enzyme Active Sites and their Reaction Mechanisms. DOI: https://doi.org/10.1016/B978-0-12-821067-3.00037-4

37.2 Physiological function

TS is the sole source of dTMP in DNA biosynthesis. Since TSase is overexpressed in tumor cells, this enzyme is an active target for the development of anticancer drugs.

37.3 Key structural features

Most of the structural and mechanistic information about TSase comes from studies on the bacterial enzymes derived from *Lactobacillus casei* and *Escherichia coli*. There is an extensive degree of structural homology between these and with other TSases (including human). The enzymes are homo-dimers with two active sites; both subunits contribute to these active sites. Five "critical residues" have been singled out as playing an essential role at the active site. For *L. casei* (corresponding *E. coli* residues in parentheses) these are Glu_{60} (Glu_{58}), Tyr_{146} (Tyr_{94}), Cys_{198} (Cys_{146}), Arg_{218} (Arg_{166}), and Asp_{221} (Asp_{169}). These five residues are shown in Fig. 37.3 in one of several structures that have

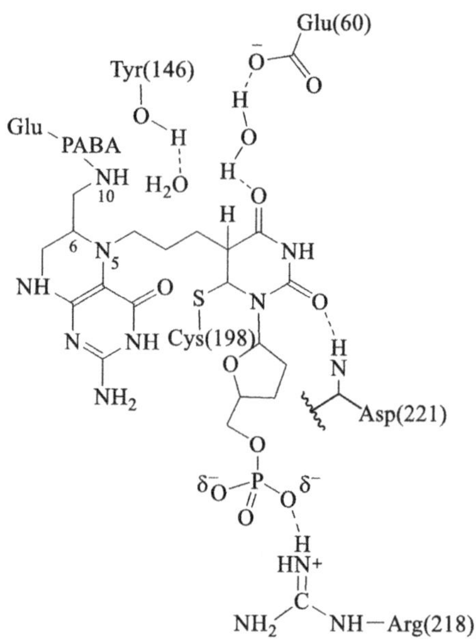

FIGURE 37.3 A structure of the *L. casei* TSase active site during the catalyzed formation of dTMP showing the five residues deemed critical for this enzyme. *(dTMP*, 2′-Deoxythymidine-5′-monophosphate. *Adapted with permission from Finer-Moore, J. S. et al.* Biochemistry ***42****, 248–256 (2003); Copyright 2003 American Chemical Society).*

been published for the active site at various stages of the reaction. In this *L. casei* structure the Cyt C-S and folate/uracil bonds have formed and the five-membered ring involving N_{10} has opened (see the mechanism outlined below).

The roles played by these residues have been a subject of considerable speculation. The function of the Cys_{198} in initiating the chemistry (see below) is without argument. Likewise, there is consensus that the positively charged Arg_{218} provides electrostatic stabilization of the thiolate anion during points in the mechanism where it detaches from the uracil ring. Several other nearby arginine residues (not shown) coordinate with the phosphate group. The role of the Glu_{60} is less direct. During the chemistry, a base is needed for proton abstraction and this function is currently assigned to a water molecule propitiously located near the target hydrogen. Glu_{60} is thought to increase the basicity of this water molecule, either directly or via a chain of hydrogen-bonded waters. The role of the base had been ascribed to Tyr_{146} but more recently, it is suggested that this residue plays a role similar to that of Arg_{218}, stabilization of the thiolate anion (by hydrogen bonding). It has been proposed, but not confirmed, that Asp_{221} serves as the acid needed to protonate the cofactor. It participates in a series of hydrogen bonds with uracil (shown), as well as with the folate and active site water (not shown).

37.4 Reaction sequence

The chemistry here is complex and proceeds in several steps. A simplified overview is shown schematically in Fig. 37.4. Step (a) shows the acid-catalyzed

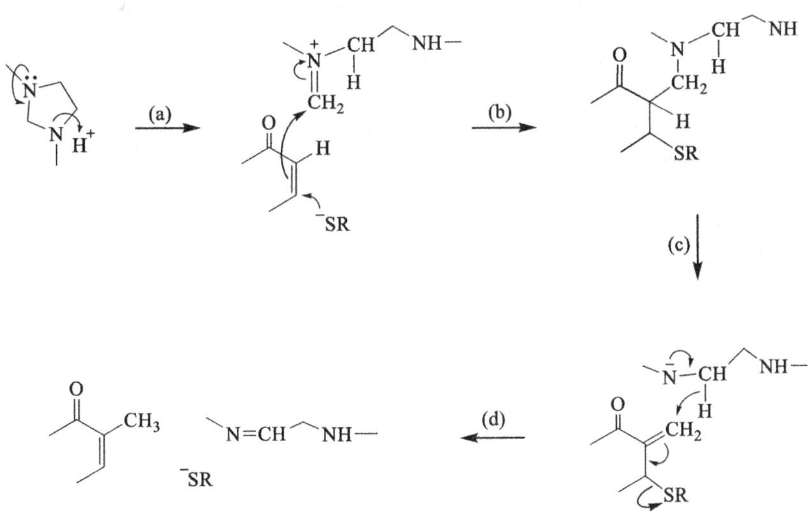

FIGURE 37.4 Overview of the chemistry catalyzed by TSase. *TSase*, Thymidylate synthase.

formation of the iminium species that sets up the molecule for the subsequent formation of a new C–C bond. That bond is generated in step (b) by the nucleophilic attack of a thiolate anion on the uracil ring. Step (c) is actually a composite of several steps detailed in Fig. 37.5. It involves a hydride ion transfer to generate the ultimate C-5 methyl group of the thymine ring. The hydride transfer is accompanied by the concomitant elimination of the catalytic thiolate group.

37.5 Detailed mechanism(s) and the roles of active site residues

It is known that formation of the ternary complex starts with the binding of dUMP to the active site, followed by binding of the cofactor. The active site closes after the cofactor is bound. A step-by-step outline of the (two) postulated mechanisms for the ensuing chemistry is given in Fig. 37.5.

The chemistry is initiated (steps a and b) by the acid-catalyzed cleavage of the N_{10}-C_{11} bond of the imidazolium ring of the MTHF. This is followed (step c) by the nucleophilic attack of the conserved Cys_{198} (*L. Casei* numbering) anion on the 5,6-double bond of the uracil ring. There has been considerable debate as to whether this attack generates an intermediate enolate anion, or whether, as drawn, the creation of the new C-C bond is concerted with C-S bond formation. Current theory and recent experimental date support the concerted process (Note the analogous role of a Cys residue in initiating the chemistry of m^5C-cytosine methyltransferase.).

The sequence of steps (d) and (e) have likewise been a subject of debate. Again, the "classical" mechanism invokes an intermediate enolate anion, but the more recent view is the expulsion, and then re-addition of the thiolate anion with concomitant cleavage of the N-CH_2 bond. The last step (f) involves a hydride anion transfer to create the 5-methyl group and thymine ring. Despite earlier attempts to identify AA residues that might serve as an acid or a base in this mechanism, the current view is that active site water, hydrogen bonded to and organized by well-conserved AA residues (Glu_{60} and Tyr_{146} are the most commonly cited; see Fig. 37.3) serve these roles.

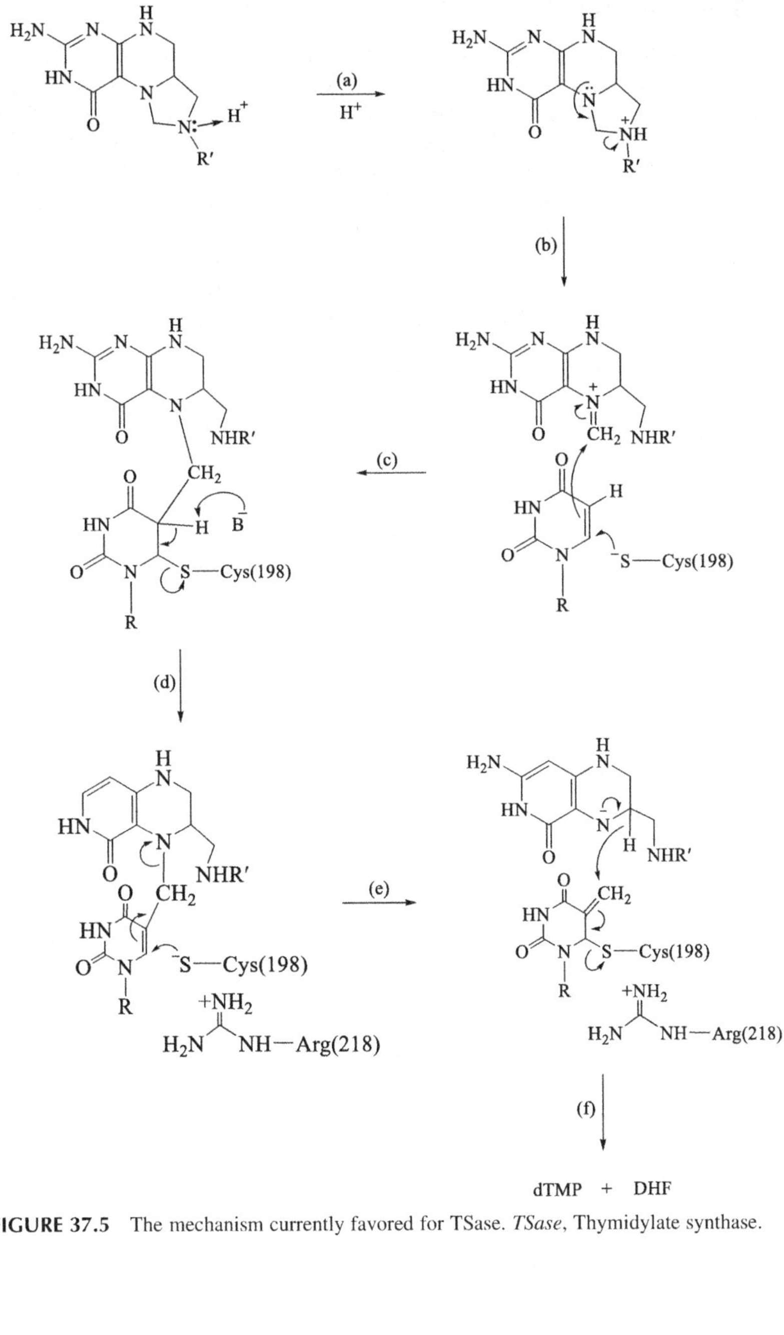

FIGURE 37.5 The mechanism currently favored for TSase. *TSase*, Thymidylate synthase.

Leading references

Pozzi, C.; Ferrari, S.; Luciani, R.; Tassone, G.; Costi, M. P.; Mangani, S. *Molecules* **24**, 1257 (2019); Kholodar, S. A.; Ghosh, A. K.; Świderek, K.; Moliner, V.; Kohen, A. *Proc. Natl. Acad. Sci. U. S. A.* **115**, 10311–10314 (2018); Finer-Moore, J. S.; Lee, T. T.; Stroud, R. M. *Biochemistry* **57**, 2786–2795 (2018); Gurevic, I.; Islam, Z.; Świderek, K.; Trepka, K.; Ghosh, A. K.; Moliner, V.; Kohen, A. *ACS Catal.* **8**, 10241–10253 (2018); Kholodar, S. A. and Kohen, A. *J. Am. Chem. Soc.* **138**, 8056–8059 (2016); Świderek, K.; Kohen, A.; Moliner, V. *Phys. Chem. Chem. Phys.* **17**, 30793–30804 (2015); Ghosh, A. K.; Islam, Z.; Krueger, J.; Abeysinghe, T.; Kohen, A. *Phys. Chem. Chem. Phys.* **17**, 30867–30875 (2015); Kaiyawet, N.; Lonsdale, R.; Rungrotmongkol, T.; Mulholland, A. J. *J. Chem. Theor. Comput.* **11**, 713–722 (2015); Finer-Moore, J. S.; Santi, D. V.; Stroud, R. M. *Biochemistry* **42**, 248–256 (2003).

Chapter 38

The 20S proteasome

38.1 The 20S proteasome

The 20S proteasome (proteasome endopeptidase complex; core particle; CP) is most commonly associated with the protease complex, the 26S proteasome (EC: 3.4.25.1), within which it functions as the site of proteolytic catalysis. The 26S proteasome is the primary source of protein degradation within eukaryotes. However, there is now ample evidence that the CP can operate independently of the "19S regulator" proteins that, when bound to the top and bottom regions of the CP, form the fully constituted 26S proteasome. Key differences between the 20S and 26S proteasomes are the dependence of the latter on prior ubiquitination of target proteins and a requirement for ATP. In combination with the 19S regulator, these allow for the unfolding of a protein, a requisite for its entry into the core of the 26S proteasome assembly. As discussed below, the 20S proteasome has no such mechanism available to it, and therefore its targets are (must be) already partially "natively" unfolded, [intrinsically disordered proteins (IDPs); intrinsically unstructured proteins (IUPs)] or have become unfolded due to aging and/or damage. The overall reaction is the hydrolysis of a peptide bond, as shown in Fig. 38.1.

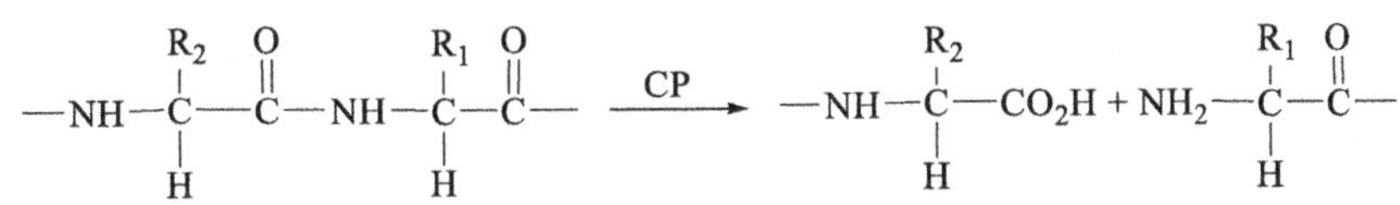

FIGURE 38.1 Overall reaction catalyzed by CP. *CP*, Core particle.

38.2 Physiological function

Proteasomes are found in all forms of life and are critical for the regulation of multiple cellular processes. They control the concentration and lifetime of numerous proteins. Assays over multiple eukaryotic cell lines indicate that 64% of cellular proteasome consists of free CP. It is found in both the cyto-

Enzyme Active Sites and their Reaction Mechanisms. DOI: https://doi.org/10.1016/B978-0-12-821067-3.00038-6

plasm and the nucleus. Proteolysis by the CP, when acting alone, specifically functions as a mechanism for the control of, and/or degradation and elimination of, proteins containing disordered regions. These include IDPs as well as proteins disordered because of stress, heat, and oxidation by reactive oxygen species. In the latter case, there is evidence that the CP is the primary source of degradation when such damage has occurred. Numerous neurological diseases are associated with misfolded, and hence aggregated, proteins. It has been suggested that the CP is important for the elimination of these proteins, and that their accumulation over time may be associated with a diminution of CP activity. An example is α-synuclein, an IUP that is highly expressed in the brain and is degraded by the CP. The accumulation and aggregation of this protein are characteristic of Parkinson's disease.

38.3 Key structural features

The CP is a barrel-like structure made up of four concentric rings stacked vertically so as to create a central channel. The N-terminal residues of the top and bottom "α" rings act as gates that control entry into the channel. They allow entry by being conformationally rearranged upon binding to a protein "activator" [PA28 (11S; REG); PA200 (Blm10)]. The middle two "β" rings contain the protease active sites. Each of the four rings is made up of seven subunits. For mammals, proteolysis involves active sites within three of the β subunits (β1, β2, and β5), with each showing unique specificity: caspase-like (post-glutamyl-peptide hydrolytic; cleavage occurs after acidic residues), trypsin-like (cleavage occurs after basic residues), and chymotrypsin-like (cleavage occurs after bulky hydrophobic residues), respectively. The active site of *T. acidophilum* bound to the inhibitor, Ac-Leu-Leu-norleucinal, is shown in Fig. 38.2.

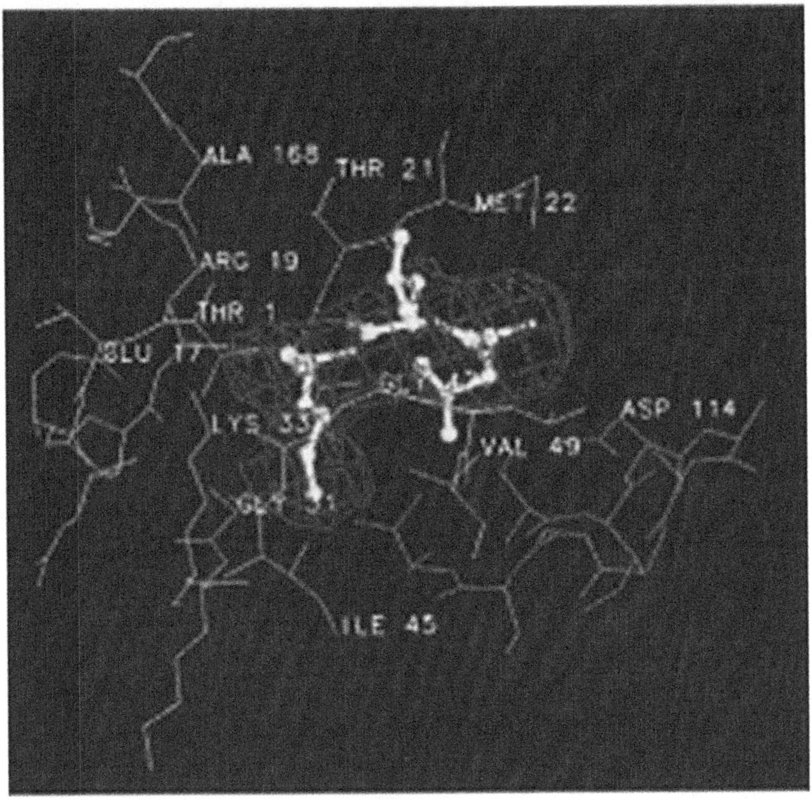

FIGURE 38.2 Active site of the 20S proteosome from *T. acidophilum* containing the inhibitor, Ac-Leu-Leu-norleucinal (calpain inhibitor I). *(Republished with permission of the American Association for the Advancement of Science from Löwe, J. et al.* Science ***268****(5210), 533–539 (1995); Copyright 1995; permission conveyed through Copyright Clearance Center, Inc.).*

The active sites are capped by N-terminal propeptides which must be removed for the CP to become functional. This occurs autocatalytically using peptides that are also involved in the proteolysis. Specifically, a catalytic triad of Lys_{33}, Thr_1, and Asp_{17} (or Glu_{17}) is found to be common to all the active sites (*Thermoplasma* CP numbering). A proposed mechanism is shown in Fig. 38.3. In step (a), the Lys_{33} amino group serves as a base to deprotonate the Thr_1 OH group. The oxyanion thus formed nucleophilically attacks the peptide linkage blocking the Thr_1 amino group. The reaction is facilitated by Asp_{17} and Arg_{19}, which interact with Lys_{33} throughout, and Gly_{47} which stabilizes the oxyanion (not shown). In step (b), the oxazolidine ring is cleaved concomitant with a series of proton transfers involving Ser_{129} that terminate in the formation of the Asp_{166} carboxylate anion. The result of this sequence is the conversion of a Thr amide into a Thr ester. Hydrolysis of the ester occurs in steps (c) and (d), wherein the liberated Thr amino group can now act as a base to activate a water molecule that is present within the active site [see step c in Fig. 38.4]. Gly_{47} again stabilizes the intermediate oxyanion.

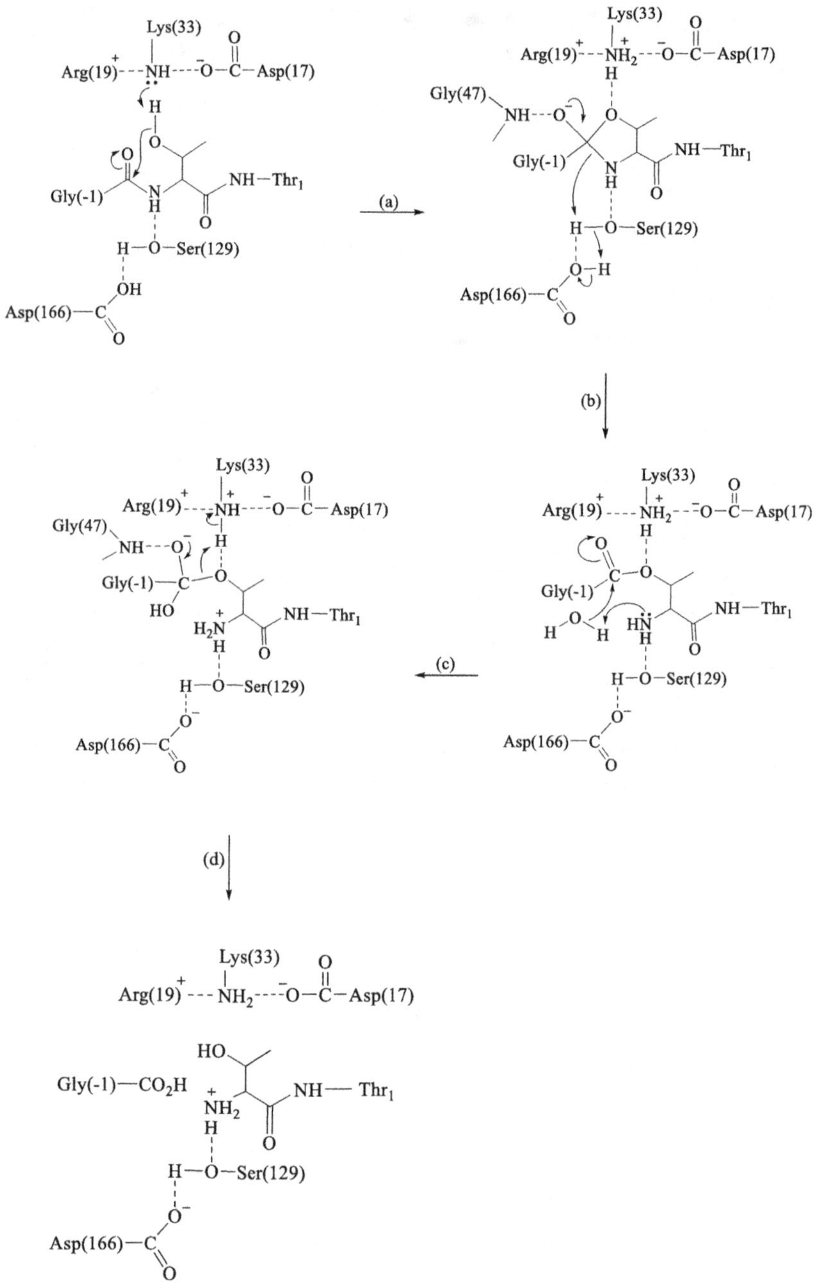

FIGURE 38.3 A proposed mechanism for the autocatalytic removal of the propeptides bound to threonine in the active site. *(Adapted by a Creative Commons License from Huber, E. M. et al.* Nat. Commun. *7, 10900–10909 (2016); courtesy Springer).*

38.4 Reaction sequence

Though complex in its details (see below), the net consequence of the sequence of steps is a base-catalyzed hydrolysis of an amide bond.

38.5 Detailed mechanism and the role of active-site residues

As noted above, the key catalytic residues are common to all three β active sites. The difference in specificity among these sites has been attributed to residues surrounding the binding pocket, especially at position 45. The basic Arg_{45} facilitates the post acidic specificity of β_1, the small Gly_{45} enlarges the pocket and helps accommodate basic P1 residues, and the apolar Met_{45} of β5 is consistent with that site's preference for hydrophobic residues.

The most recent proposed mechanism for proteolysis is shown in Fig. 38.4. The CP is a threonine protease (an enzyme that uses the alcohol of an N-terminal threonine as a nucleophile during catalysis). Step (a) involves deprotonation of the Thr_1 hydroxyl group by the Lys_{33} amino group, with the proton transfer facilitated by Asp_{17} (and Arg_{19}) (as in Fig. 38.1). This proposal reverses the roles of Lys_{33} and the terminal amino group of Thr_1 assigned in earlier mechanisms. The oxyanion thus generated attacks the target amide carbonyl functionality. The resulting tetrahedral intermediate is again stabilized by the oxyanion hole created by Gly_{47}. Cleavage and proton transfer (step b) generate a Thr_1 ester which is hydrolyzed (steps c and d) by a water molecule made basic by the nearby Thr_1 amino group.

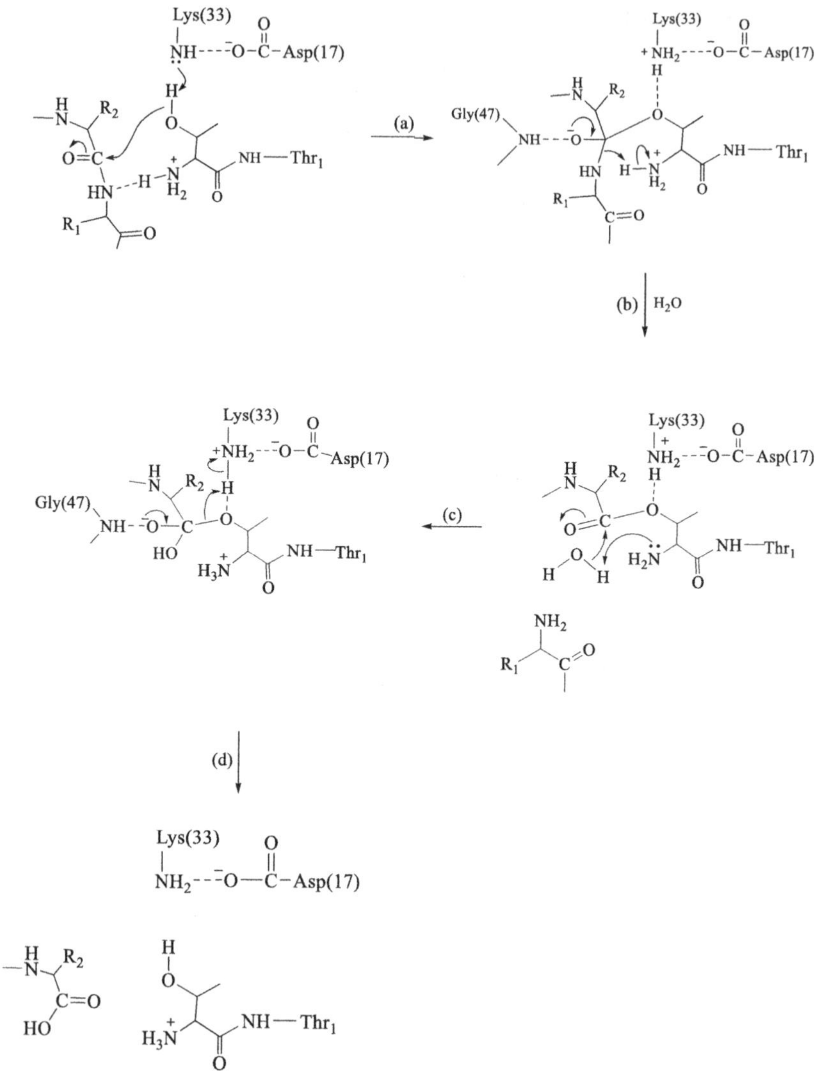

FIGURE 38.4 Mechanism for proteolysis of a target protein by the 20S CP. From Huber, E. M. et al. *Nat. Commun.* **7**, 10900–10909 (2016), see Fig. 38.3.

Leading references

Majumder, P. and Baumeister, W. *Biol. Chem.* **40**, 183–199 (2020); Hodošček, M. and Elghobashi-Meinhardt, N. *Int. J. Mol. Sci.* **19**, 3858 (2018); Budenholzer, L.; Cheng, C. L.; Li, Y.; Hochstrasser, M. *J. Mol. Biol.* **429**, 3500–3524 (2017); Jones, C. L.; Njomen, E.; Sjögren, B.; Dexheimer, T. S.; Tepe, J. J. *ACS Chem. Biol.* **12**, 2240–2247 (2017); Becker, S. H. and Darwin, K. H. *Microbiol. Mol. Biol. Rev.* **81**, e00036-16 (2017); Huber, E. M.; Heinemeyer, W.;

Li, X.; Arendt, C. S.; Hochstrasser, M.; Groll, M. *Nat. Commun.* **7**, 10900–10909 (2016); Raynes, R.; Pomatto, L. C. D.; Davies, K. J. A. *Mol. Aspects Med.* **50**, 41–55 (2016); Erales, J. and Coffino, P. *Biochim. Biophys. Acta* **1843**, 216–221 (2014); Marques, A. J.; Palinimurugan, R.; Matias, A. C.; Ramos, P. C.; Dohmen, R. J. *Chem. Rev.* **109**, 1509–1536 (2009); Wlodawer, A. *Structure* **3**, 417–420 (1995); Löwe, J.; Stock, D.; Jap, B.; Zwickl, P.; Baumeister, W.; Huber, R. *Science* **268**, 533–539 (1995).

Chapter 39

Uracil-DNA glycosylase

39.1 Uracil-DNA glycosylase

Uracil-DNA glycosylase [EC 3.2.2.3; uracil-N glycosylase; UDG; UNG; hUNG (human)] is a repair enzyme that removes (excises) dU from DNA. DNA repair enzymes are ubiquitous and essential to all forms of life (see also Chapter 13: Deoxyribodipyrimidine photolyase). UDG is one of six glycosylase "superfamilies." hUNG exists in both the mitochondria (UNG1) and the nucleus (UNG2). The dU lesions in DNA are formed from both the naturally occurring, hydrolytic deamination of cytosine (see Fig. 39.1) and

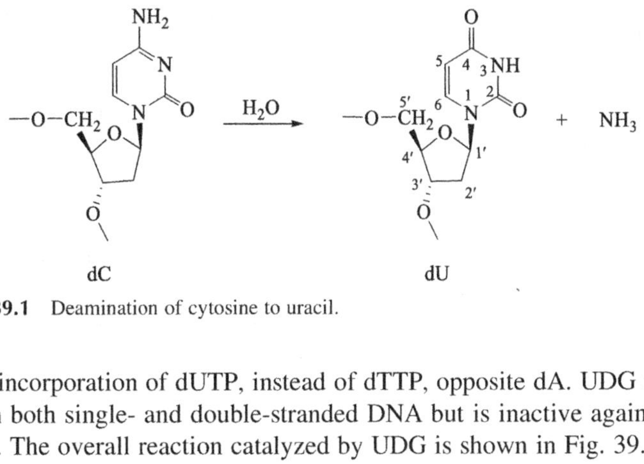

FIGURE 39.1 Deamination of cytosine to uracil.

the mis-incorporation of dUTP, instead of dTTP, opposite dA. UDG removes dU from both single- and double-stranded DNA but is inactive against the U of RNA. The overall reaction catalyzed by UDG is shown in Fig. 39.2.

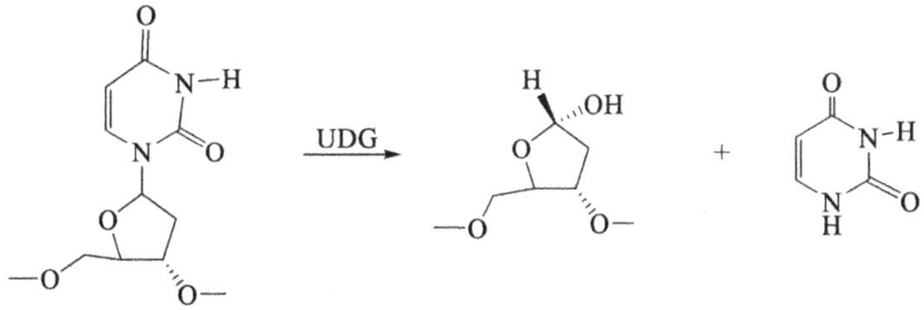

FIGURE 39.2 Overall reaction catalyzed by UDG. *UDG*, Uracil-DNA glycosylase.

Enzyme Active Sites and their Reaction Mechanisms. DOI: https://doi.org/10.1016/B978-0-12-821067-3.00039-8

39.2 Physiological function

The creation of dU opposite dG (or mis-incorporation in place of dT opposite dA) is highly mutagenic. UDG initiates a repair pathway by cleavage of the N-glycosidic bond to form an apyrimidinic (AP) site. Base excision repair (BER) then follows, wherein dC (or dT) is inserted by a sequence in which (1) an AP-endonuclease (APE1) hydrolyses the deoxyribose 5′-phosphate linkage, (2) a DNA polymerase (Pol β) further hydrolyses the deoxyribose 3′-phosphate linkage and inserts dC (dT), and (3) a DNA ligase (Lig1 in nuclei; Lig3 in mitochondria) seals the chain. The process is called "short-patch BER" or "single-nucleotide BER" (see Fig. 39.3). See Chapter 13, Deoxyribodipyrimidine photolyase for "nucleotide excision repair."

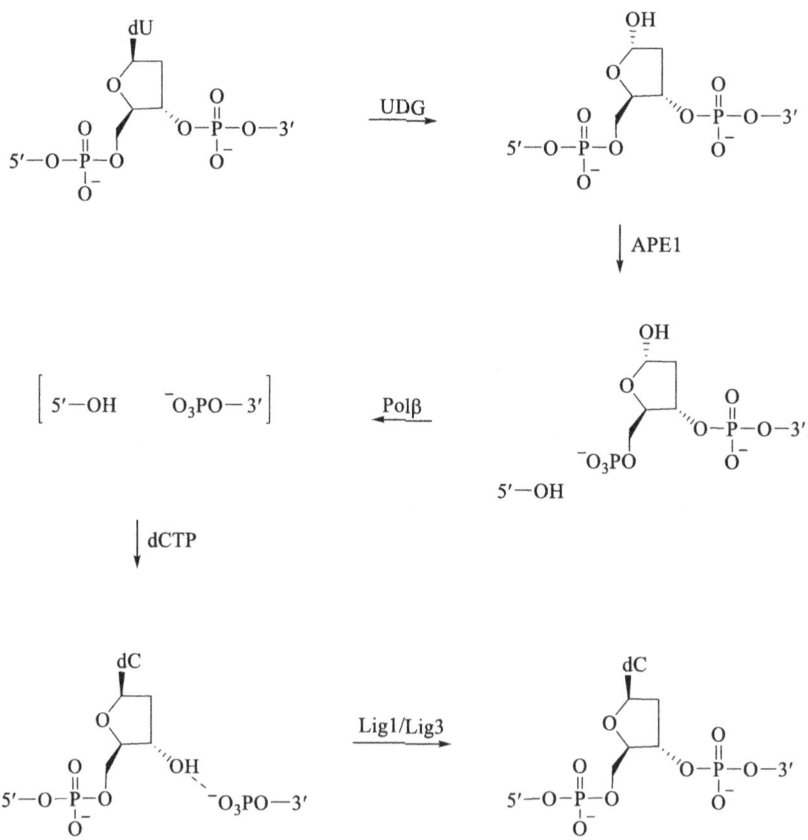

FIGURE 39.3 Short-patch BER illustrated by replacement of dU by dC. *BER*, Base excision repair.

39.3 Key structural features

A representation of the binding pocket for hUNG is shown in Fig. 39.4. Of particular note are the tyrosine and asparagine residues which help discriminate against thymine (methyl collision at C5) and cytosine (hydrogen bonding pattern), respectively.

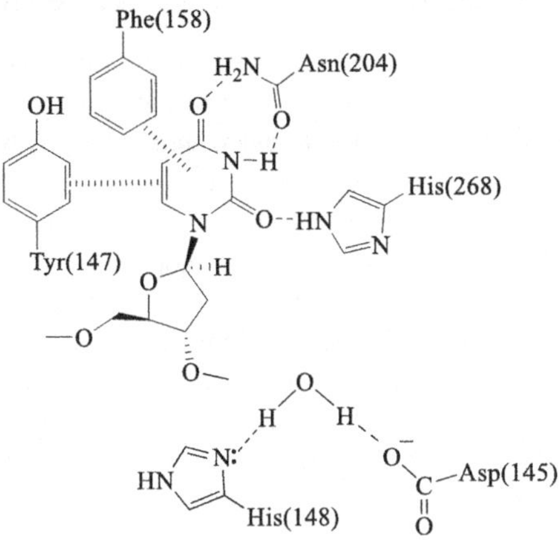

FIGURE 39.4 Key residues within the *h*UNG dU binding pocket prior to cleavage of the glycosidic bond. *(Adapted with permission from Przybylski, J. L.; Wetmore, S. D.* Biochemistry *50, 4218–4227 (2011); Copyright 2011 American Chemical Society).*

39.4 Reaction sequence

The chemistry is best described as a classical S_N1 (more recently, $D_N \times A_N$) nucleophilic displacement reaction. This reaction, in which the dissociation of the leaving group is rate-determining, creates a carbenium ion intermediate that is stabilized by the lone-pair electrons on the adjacent oxygen (see Fig. 39.5). Note that, in this case, the uracil moiety leaves as an anion, since it is insufficiently basic to be protonated prior to the cleavage. That creates a challenge for the enzyme which must stabilize

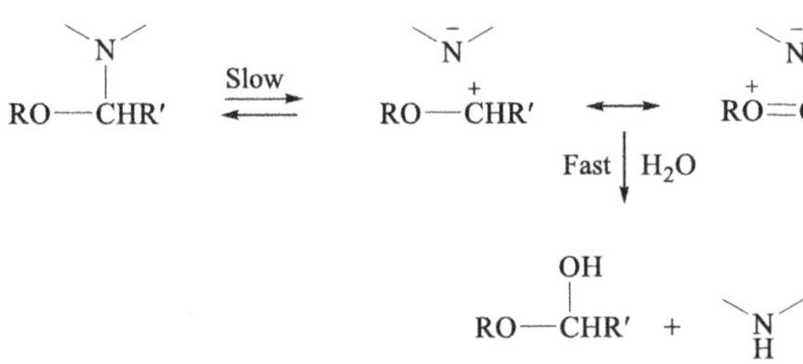

FIGURE 39.5 The reaction sequence for glycosylase cleavage involving the S_N1 mechanism.

both the developing positively charged carbenium ion and the developing uracil anion in the rate-determining transition state. Note that the C1′ reaction center may be viewed as the alkyl ether of a carbinol amine converting to a hemiacetal.

39.5 Detailed mechanism and role of active-site residues

There is overwhelming evidence that base excision only occurs after the dU has been "flipped out" of the DNA helix and into the recognition pocket of UDG. Exactly how the relatively rare dU is found within the DNA has been a matter of much debate and speculation. UDG nonspecifically binds to the minor groove of DNA and must then "scan" for a site of damage. It is thought that it does so by sliding across the DNA surface followed by "hopping" to an adjacent region to repeat the process. There are two views regarding the "recognition" of dU by the enzyme: (1) that each of the DNA bases, including, dU, is naturally and spontaneously placed into a partially extrahelical position prior to enzyme recognition and (2) that the enzyme induces a "kinking" of the DNA during the "search" process, "interrogates" bases exposed by this process, and facilitates the flipping of a dU out of the helix into a recognition site. What is undebatable is that once a dU is detected by the UDG recognition pocket, a highly conserved leucine residue is inserted to fill the void left by the reoriented base and help stabilize the local sequence of base pairs.

The most recent mechanistic proposal is outlined in Fig. 39.6. In step (a), the N-glycosidic bond is cleaved to form the oxacarbenium ion and the uracilate anion. The proximity of these two moieties allows for some stabilization of the "ion-pair"; phosphate groups are also thought to contribute to the stabilization of the cation. The uracil N1 is normally insufficiently acidic for this cleavage to be energetically feasible, but His_{268} plays a key role in reducing the N1 pKa. Note that both of the ionic products are stabilized by additional resonance forms that delocalize the charges. In step (b), His_{148} deprotonated a nearby water molecule which then reacts with the cation to stereospecifically hydrate the deoxyribose ring. Note that Asp_{145} has been alternatively invoked as the basic moiety in this step. The proton is transferred from the His_{148} to the aspartate anion to complete the reaction. Interestingly, the enzymatic reaction is complete with the uracil anion yet unprotonated.

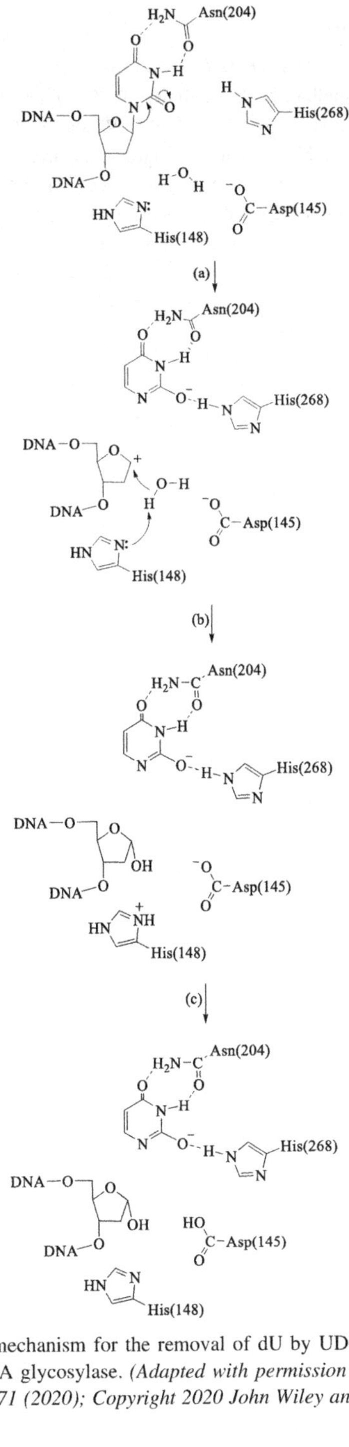

FIGURE 39.6 Detailed mechanism for the removal of dU by UDG and the role of active-site residues. *UDG*, Uracil-DNA glycosylase. *(Adapted with permission from Kaur, R. et al.* WIREs Comput. Mol. Sci. ***10****, e1471 (2020); Copyright 2020 John Wiley and Sons).*

Leading references

Kaur, R.; Nikkel, D.; Wetmore, S. D. *WIREs Comput. Mol. Sci.* **10**, e1471 (2020); Mullins, E. A.; Ridriguez, A. A.; Bradley, N. P.; Eichman, B. F. *Trends Biochem. Sci.* **44**, 765–781 (2019); Beard, W. A.; Horton, J. K.; Prasad, R.; Wilson, S. H. *Annu. Rev. Biochem.* **88**, 137–162 (2019); Dizdaroglu, M.; Coskun, E.; Jaruga, P. *Mutat. Res.* **771**, 99–127 (2017); Drohat, A. C. and Coey, C. T. *Chem. Rev.* **116**, 12711–12729 (2016); Schormann, N.; Ricciardi, R.; Chattopadhyay, D. *Protein Sci.* **23**, 1667–1685 (2014); Przybylski, J. L. and Wetmore, S. D. *Biochemistry* **50**, 4218–4227 (2011); Zharkov, D. O.; Mechetin, G. V.; Nevinsky, G. A. *Mutat. Res.* **685**, 11–20 (2010); Friedman, J. I. and Stivers, J. *Biochemistry* **49**, 4957–4967 (2010); Bianchet, M. A.; Seiple, L. A.; Jiang, Y. L.; Ichikawa, Y.; Amzel, L. M.; Stivers, J. T. *Biochemistry* **42**, 12455–12460 (2003).

Chapter 40

Vanadium-dependent chloroperoxidase

40.1 Vanadium chloroperoxidase

Vanadium chloroperoxidase (EC 1.11.1.10; V-CPO; VCPO) is one of a family of vanadium haloperoxidases (VHPO) that can chlorinate, brominate, and iodinate organic substrates. They are characterized by the common use of hydrogen peroxide to oxidize halide anions to electrophilic "X^+" species. These vanadium-dependent peroxidases are, in turn, one member of a larger family of halogenases that employ multiple mechanisms and cofactors to halogenate substrates. The overall reaction initiated by VCPO is shown in Fig. 40.1. The vanadium is strictly catalytic and is in the +5 oxidation state throughout the course of a multistep reaction mechanism (see below).

$$Cl^- + H_2O_2 + RH + H^+ \longrightarrow R-Cl + 2H_2O$$

FIGURE 40.1 The overall chemistry catalyzed by vanadium chloroperoxidase.

40.2 Physiological function

VCPO is found in marine algae, terrestrial fungi, and bacteria. The enzyme is thought to serve a protective function by producing hypochlorous acid (HOCl). VCPO also participates in the biosynthesis of chlorinated natural products, including terpenoids and antibiotics.

40.3 Key structural features

The active site for VCPOs is conserved throughout the family. An X-ray structure for the fungus *Colletes inaequalis* shows it to lie at the bottom of a channel, one side of which is lined primarily with AA residues that are polar and primarily with those that are hydrophobic. The vanadium atom is in the +5 oxidation state and is bound to five ligands in a trigonal bipyramid arrangement. These include an apical His_{496} through which it is anchored to the enzyme at the bottom of the channel. In addition, there are four bound oxygen atoms—one apical and three equatorial. Two protons are associated with the vanadate. Calculations place them

Enzyme Active Sites and their Reaction Mechanisms. DOI: https://doi.org/10.1016/B978-0-12-821067-3.00040-4

both on the apical oxygen in the lowest energy structure, with each equatorial oxygen bearing a negative charge. The latter are stabilized by ion pairing and hydrogen bonding interactions with active-site residues (cf. Fig. 40.2). However, a resting state of the enzyme that has one hydrogen on the apical oxygen and one on an equatorial oxygen is virtually isoenergetic with this structure. In fact, there is X-ray evidence for a mixture of these two structures; the mechanism below starts with one proton each on the apical and an equatorial oxygen.

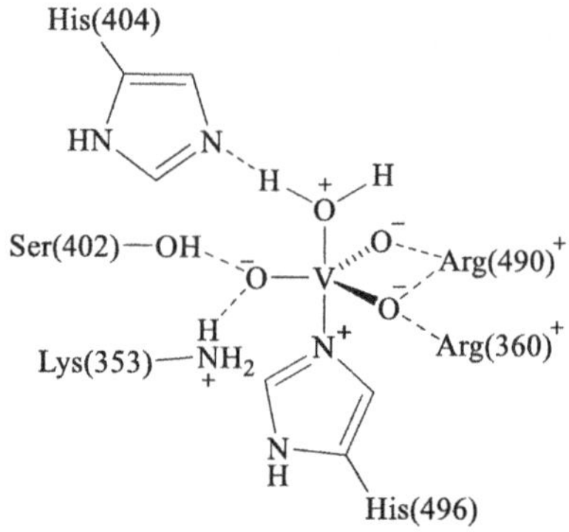

FIGURE 40.2 The active site of *C. inaequalis* with a diprotonated apical oxygen. *(From LeBlanc, C. et al.* Coord. Chem. Rev. ***301–302****, 134–146 (2015)).*

There is also evidence for solvent water molecules at the active site that are not shown. A key intermediate in the mechanism for VCPO (see below) is a peroxo species, for which an X-ray structure has also been obtained. The vanadium atom remains bound to His_{496} and is stabilized by several of the residues seen in Fig. 40.2, that is, Lys_{353}, Arg_{350}, and Arg_{490}. The peroxide oxygen atoms are hydrogen bonded to the amide hydrogen of Gly_{403}.

Two active-site residues, Trp_{350} and Phe_{397}, are important in the binding of the chloride anion. In fact, replacing the corresponding residue at position 350 in the bromine analog, VBPO, with a Trp has been shown to increase affinity for the chloride anion. Less is known about the potential binding of a target substrate near the active site. It has generally been assumed that the hypochlorous acid leaves the active site upon its formation. A two-phase system has been reported in which the brominating agent from a VBPO in water is transferred to an organic solvent whence it reacts with a target substrate. Nevertheless, there are reports of regio- and stereospecificity in enzymatic halogenations that differ from organic chemical model reactions, thus implying a close encounter with the substrate in a stereocontrolled manner in these instances.

40.4 Reaction sequence

The reaction shown in Fig. 40.1 occurs in a multistep sequence, the details of which are not all well-characterized. In particular, there is no unanimity in the literature on the number and location of protons on several intermediates. One proposed sequence, supported by computation, is shown in Fig. 40.3. Here, the first step (a) involves attack by hydrogen peroxide on the mono-protonated vanadate to generate a vanadium peroxide intermediate. (As noted above, a peroxo intermediate, absent in the two hydrogen atoms, has been characterized by X-ray analysis). This is attacked by the chloride anion (step b) to form a vanadium hypochloride, which releases the chlorinating agent, hypochlorous acid (step c).

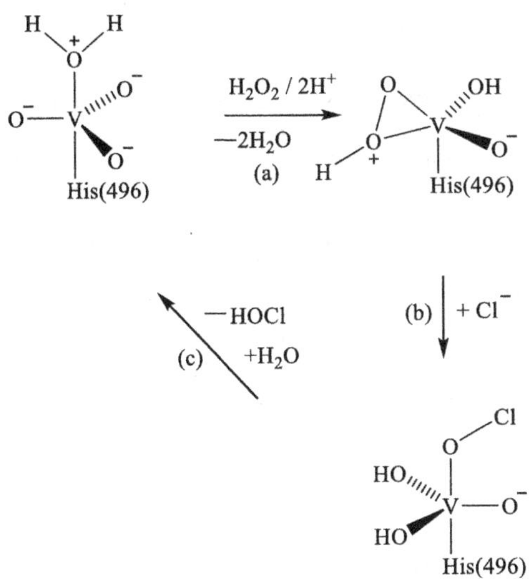

FIGURE 40.3 A proposed reaction sequence for the formation of hypochlorous acid by VCPO.

40.5 Detailed mechanism and the role of active-site residues

Though the peroxo intermediate is well-characterized, the individual steps leading to its formation, its subsequent attack by the chloride anion, and the ejection of hypochlorous acid, are not yet fully elucidated. The mechanism shown below in Fig. 40.4 is therefore a composite of several proposals, each of which differs slightly in the number and placement of protons during the reaction. It draws heavily from Kravitz et al. (2005) and Messerschmidt et al. (1999) (see below). It is known that the initiating attack by hydrogen peroxide on the vanadate requires that His_{404} be deprotonated, and thus it has been assumed that this residue facilitates the deprotonation of the H_2O_2 by the apical oxygen. The hydroperoxide anion then replaces the apical aquo ligand, either by an "associative" (i.e., addition of the peroxide followed by elimination of water") or a "dissociative" (i.e., elimination of the water followed by addition of the peroxide) mechanism.

Fig. 40.4 begins with the entry of hydrogen peroxide into the active site (step a). In step (b), the apical oxygen deprotonates the H_2O_2, with the resultant hydroperoxide anion then bonding to the vanadium atom by displacement of the apical water molecule (step c). Intramolecular displacement of hydroxide anion in step (d) is followed by protonation of one of the peroxo oxygen atoms by Lys_{353} (step e). This activates the O—O bond for cleavage by the nucleophilic attack by the chloride anion (step f). The apical OCl linkage is protonated by a solvent water molecule (step g) and hypochlorous acid is eliminated concomitant with the reformation of the diprotonated vanadate species.

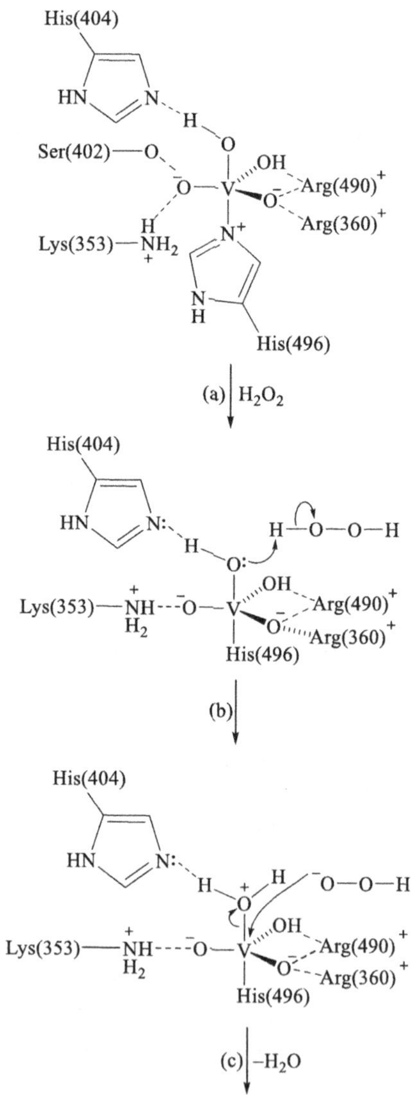

FIGURE 40.4 A detailed mechanism with the role of active-site residues.

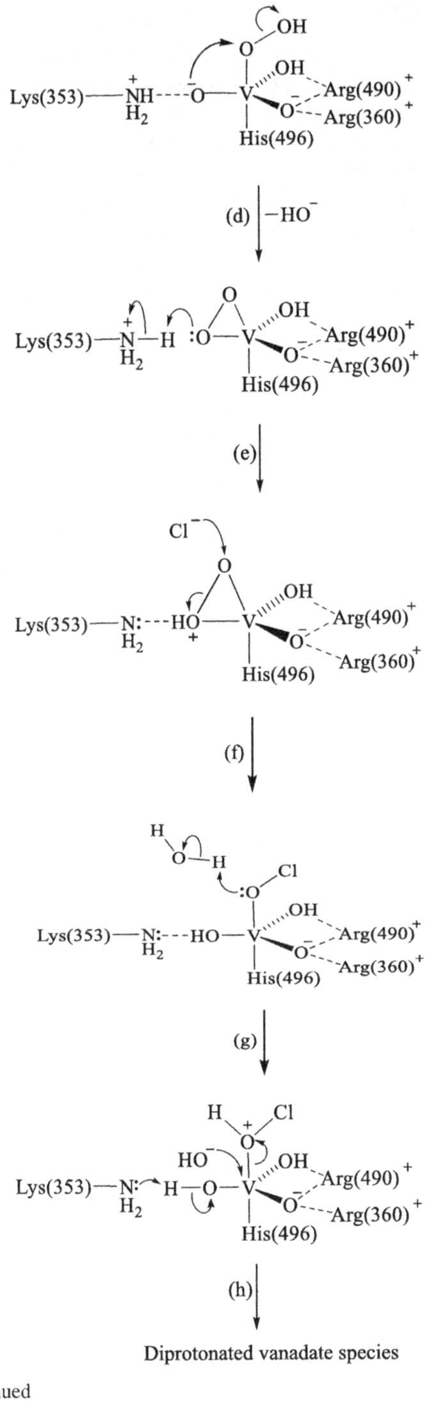

FIGURE 40.4 continued

Leading references

Agarwal, V.; Miles, Z. D.; Winter, J. M.; Eustáquio, A. S.; El Gamal, A.; Moore, B. S. *Chem. Rev.* **117**, 5619–5714 (2017); Leblanc, C.; Vilter, H.; Fournier, J.-B.; Delage, L.; Rebuffet, E.; Michel, G.; Solari, P. L.; Felters, M. C.; Czjek, M. *Coord. Chem. Rev.* **301–302**, 134–146 (2015); Conte, V. and Floris, B. O. *Inorg. Chim Acta* **363**, 1935–1946 (2010); Winter, J. M. and Moore, B. S. *J. Biol. Chem.* **284**, 18577–18581 (2009); Hasan, Z.; Renirie, R.; Kerkman, R.; Ruijssenaars, H. J.; Hartog, A. F.; Wever, R. *J. Biol. Chem.* **281**, 9738–9744 (2006); Kravitz, J. Y.; Pecoraro, V.; Carlson, H. A. *J. Chem. Theory Comput.* **1**, 1265–1274 (2005); Zampella, G.; Fantucci, P.; Pecoraro, V.; De Gioia, L. *J. Am. Chem. Soc.* **127**, 953–960 (2005); Macedo-Ribeiro, S.; Wieger, H.; Renirie, R.; Wever, R.; Messerschmidt, A. *J. Biol. Inorg. Chem.* **4**, 209–219 (1999); Hemrika, W.; Renirie, R.; Macedo-Ribeiro, S.; Messerschmidt, A.; Wever, R. *J. Biol. Chem.* **274**, 23820–23827 (1999).

Index

Note: Page numbers followed by "*f*" refer to figures.

B

C

D

E

F

G

H

I

K

L

M

N

R

S

T

U

V

W

Z

Made in the USA
Monee, IL
13 September 2026

63138390R20155